AF324861

Food Safety Management in China

A Perspective from Food Quality Control System

Food Safety
Management in China

A Perspective from Food Quality Control System

Jiehong Zhou • Shaosheng Jin
Zhejiang University, China

Published by

World Scientific Publishing Co. Pte. Ltd.
5 Toh Tuck Link, Singapore 596224
USA office: 27 Warren Street, Suite 401-402, Hackensack, NJ 07601
UK office: 57 Shelton Street, Covent Garden, London WC2H 9HE

and

Zhejiang University Press
No. 148, Tianmushan Road
Xixi Campus
Hangzhou 310028, China

British Library Cataloguing-in-Publication Data
A catalogue record for this book is available from the British Library.

This edition is jointly published by World Scientific Publishing Co. Pte. Ltd. and Zhejiang University Press. This edition is distributed worldwide by World Scientific Publishing Co. Pte. Ltd., except China.

FOOD SAFETY MANAGEMENT IN CHINA
A Perspective from Food Quality Control System

ISBN 978-981-4447-75-1

In-house Editor: Lixia Chen

Typeset by Zhejiang University Press
Email: 461505327@qq.com

Printed in China.

Preface

Food quality and safety is the basic necessity of human's survival. It is an important symbol of the economic development and people's living conditions in a country. At present, being in a transitional period, China has been experiencing a series of food scandals involving dyed steamed buns and lean meat powder residues in pork meat, which raised wide public concern about domestic food safety. Subsequently, food safety bonding by finance security, food security, energy safety, and ecological safety makes up the Chinese national safety system, proposed at the 21st conference on June 29, 2011.

In recent years, the Chinese government has invested a large amount of human resources and capital resources to accelerate the treatment of food safety problems. The government has also taken various measures to strengthen oversight and law enforcement to comprehensively raise the capacity to ensure food safety and quality. Yet the target of food safety and quality system in China is not being met. A large gap in development exists among endemic industries. The food supply chain is relatively long. The majority of food producers and operators are on a small-scale. In addition, some practitioners lack social responsibility for public health. All these characteristics constitute a specific and unique environment for China's food safety.

Hence, design for a surveillance mechanism for Chinese food safety management and control should use the experiences of developed countries as references but not using everything unconditionally for future development. Instead, the Chinese government must launch a number of treatments in combination with agricultural industrialization and standardization that will fundamentally provide an effective solution to problems mentioned above.

Overall, this academic book takes vegetables, pork products and aquatic products — important agricultural products in China, as research objectives. We applied scientific methods and analytical tools, and combined these with investigation and case studies to explore a long-term mechanism for setting up Chinese food quality and safety management, and hope to provide empirical evidence for scientific management decision making for the Ministry of Agriculture and other relevant government sectors. Also, a comprehensive introduction to China's agriculture industry will add a new practical example to other countries, particularly, as an experience that can be considered as a reference for developing countries aiming at perfect food safety management.

This work is an academic monograph hosted by Professor Jiehong Zhou and Associate Professor Shaosheng Jin at the Center for Agricultural and Rural Development, Zhejiang University. Special thanks go to agricultural and economics graduate students including Juntao Ye, Zhen Yan, Kai Li, Qingyu Liu, Shidu Zhang and Yuan Wang. Most of the work involving the questionnaire survey, data collection and analysis, and much of the legwork was accomplished by them. We also greatly appreciate those departments and parties that have supported us during the investigation and data collection.

Food quality and safety management is a complex and systematic project, with continuous development requiring ongoing research, it is difficult for us to cover all aspects of food safety. We are limited to the authors' knowledge. We greatly appreciate all experts, researchers and readers to point out any mistakes or inadequacies that may exist.

Jiehong Zhou
Hangzhou, China
September, 2012

Acknowledgement

The research is supported by:

1) The National Natural Science Foundation of China (NNSFC-70903058, Research on the reasons why small and medium size food enterprises are not implementing HACCP, their willingness to pay for HACCP and policy implications);

2) The Zhejiang Provincial Natural Science Foundation (Z12G030016, Research on the persistent effect mechanism of agri-food safety management underlying environmental sustainable development in Zhejiang);

3) The Qianjiang Talents Project (Type C) of Zhejiang Province (QJC1102002, Research on the construction of agri-food safety traceability system in Zhejiang: A Consumer Perspective) and

4) The Project of National New Countryside Construction and Development, the "985 Project" of Zhejiang University.

Contents

Overview of Food Safety Management in China

1.1 Changes in the Focus of Food Safety in China

The concept of food safety has experienced a change from a quantity-oriented definition to a quality-oriented definition. In the World Food Conference held in 1974, the UN Food and Agriculture Organization (FAO) defined food safety as: under any circumstances, all people have access to adequate food necessary for healthy survival. By the 1980s, as the focus of food safety study had been changed from the security of aggregate food supply to the structure of food supply and consumer demand, more and more attention had been focused on food quality safety. In 2003, the World Health Organization (WHO) defined food safety as a public health problem and that poisonous and harmful substances in food affect human health.

Currently, most scholars tend to divide food safety into two levels: the security of food quantity (Food Security) and the safety of food quality (Food Safety). Food Security is mainly related to the security of aggregate food supply and the structure of food supply, while Food Safety is mainly associated with the quality of food and health security.

In China, due to the limited amount of arable land and the huge population pressure, food security has been given greater concern for a long period of time, and the grain supply, which is the basis of food supply, has always been the focus of policies and related research. Over the past decade, with the steady growth in grain output, especially the yield increase for seven consecutive years from 2003 and the steady aggregate yield of more than 500 million tons for five consecutive years after 2007, China's grain supply and demand falls into a situation of tight balance, obvious structural contradiction, and appropriate imports, and the concern of aggregate food supply has been slowly transferred to the structure and quality of food supply. In 2002, the Central Rural Work Conference made it clear that the task of agriculture and food industry development should be upgraded from food supply security to food quality safety. With the rapid economic development and increase in per capita income since the reform and opening up, the consumption level and consumption structure of urban and rural residents in China took on the following characteristics: first, despite the existence of urban-rural differences, the overall consumption level has increased continuously; second, the consumption structure has been diversified, instead of a grain-based one, which mainly finds expression in the steady decrease in the consumption of staple foods and in the steady increase in non-staple foodstuff consumption; third, with the enhancement of people's concern about their own health, the rural and urban residents are showing increasing solicitude for food safety. People pay much closer attention to the quality of food and have put forward higher requirements for the quality and sanitation status of food processing. To sum up, we believe that, currently, the food safety issues in China have entered a stage that highlights both food security and food safety, while the food safety concern is more prominent. Therefore, this chapter will focus on the analysis of China's food safety issues.

1.2 China's Food Safety: Understanding Based on the Perspective of Non-Traditional Security

After the 1990s, the concept of "non-traditional security" has been given more and more attention. The so-called non-traditional security is naturally opposite of the traditional security (political security and military security). However, it is very difficult to precisely define this term because, on the one hand, it covers all the threats and dangers for the survival and development of human society as a whole in a wide range of fields, like the economy, ecology, culture, and information, with the exception of politics and the military world (Yu *et al.*, 2006), and on the other hand, the characteristics and focus of non-traditional security issues of different countries vary enormously. But, it is generally believed that non-traditional security is a generalized definition of security and it is the expansion of traditional security theory, generally including economic security, environmental security, ecological security, cultural security, and information security. In essence, we believe that the core of non-traditional security is human security which touches upon the various factors directly posing a threat to the security of human beings in the real world. According to the elaboration of the United Nations Development Programme[1], human security includes two aspects: the security from the threat of long-term factors such as starvation and food-borne illness as well as the protection from unexpected damage in daily lives. Yu *et al.* (2006) listed seven major elements necessary for human security: economic security, food security, health security, environmental security, personal security, community security, and political security. Food security is obviously related to the safety of human beings and interacts with economic security and health security. Therefore, it is essential to re-examine food safety issues from the perspective of non-traditional security.

Firstly, food safety falls into the category of non-traditional security. We previously mentioned that the focus of China's food safety issues has gradual-

---

[1] Human Development Report (1994). United Nations. New York: United Nations Development Programme.

ly expanded from quantity security to quality security. But whether it is quantity security or quality security, it is closely linked with the core of non-traditional security — human security. Viewed from the perspective of quantity security, grain security is not only one of the core issues of food safety, but also a very important element among the seven major elements for human security. In addition, the grain security issues are intertwined with issues of economic security, community security and political security. Viewed from the perspective of quality security, grain security is directly related to health security, environmental security, and personal security. Seen from the impact of food safety issues, it is more obvious: (1) viewed from the microscopic perspective, that food safety issues directly affect the nutritional status and physical health of residents. (2) Viewed from the industry perspective, the generation of food safety issues has a direct relationship with the current mode of economic growth. Such an extensive mode of growth, on the one hand, resulted in a waste of resources and damage to the environment and thus a threat to the country's ecological security and environmental security; on the other hand, because of the reaction within the environment — mainly a variety of pollution, the quality of food was adversely affected, resulting in a low level of food quality. In addition, the occurrence of food safety incidents also allowed consumers to decrease their trust in food companies, which not only raised the management costs of the government, but also hindered the development of new food markets, like organic food, which is based on a credit mechanism. (3) Viewed from the perspective of the national level, food safety issues have a direct impact on the national economic security, social stability, and public confidence in the government. (4) Viewed from its international impact, the development of economic globalization connects all countries more closely in the same world market, and thus a country's food safety problems can be easily extended to other countries through trade mechanisms, resulting in global food safety crisis and even political disputes. European mad cow disease is a typical example. In short, due to the potential of causes, the proliferation of transference, the comprehensiveness of governance, and the universality and severity of influence, food safety issues have obvious characteristics of non-traditional security. Thus, we believe that China's food safety issues can be classified in the category of non-traditional security.

Secondly, research on food safety issues needs to draw lessons from the concept of non-traditional security. Between non-traditional security and traditional security, there is a big difference in their security concept, security sources, security subject, security focus, and security maintenance. Specifically, non-traditional security refers to a security concept of excellent state co-existence, and it includes not only national security, but also human security and social security. With respect to the security subject, it includes, apart from the state behavioral agent, a wider range of non-state behavioral agents, and thus the security maintenance involves the participation of all people. Viewed from the reasons for the occurrence of food safety issues, especially food quality safety issues, in addition to the imperfections of the national governance system, the inadequate technical support of food safety governance as well as producers' subjective and intentional violation of laws in the food production process should not be ignored; thus, the sources of food safety issues have much uncertainty, which is similar to non-traditional security. The governance system for food safety issues should be a system involved in the participation of multi-subjects, but the current governance system for food safety issues is more of a government-led regulatory system. In this system, the various behavioral subjects in the food supply chain are only the objects of governance and receivers of policies and they passively participate in the governance process of food safety issues; therefore, the system cannot effectively play a role. According to the concept of non-traditional security, the governance system of food safety issues should be a regulatory system, in which the government plays a leading role, in which the subjects of the supply chain, the main third-party agencies, and the media participate, in which information can be exchanged, and in which smooth communication between different levels can be achieved. Either viewed from relevant foreign experience or from the effect of current supervision of food safety issues, the latter is an ideal food safety supervision approach. In addition, enlightenment of the theory of non-traditional security for the supervision of food safety issues also lies in the premise and content of security maintenance.

The theoretical study of food safety issues typically begins with the analysis of food attributes. In fact, the so-called food safety consists of the food quality safety attributes, which can be subdivided into security attributes, including food problems that may cause damage to human health, food nutrition, as well as food quality, etc. However, regardless of what kind of quality safety attributes are studied, the related information is asymmetric to different extents. Therefore, we usually believe that the root of food safety issues is the problem of asymmetric information, and thus the governance mechanism of food safety issues should focus on how to eliminate the problem of asymmetric information. To solve this problem, scholars have performed a lot of research and argumentation from the perspective of economics and public management, and economic analysis consisting of three different approaches, namely, information economics, welfare economics, and property rights theory.

1.3 China's Food Safety Supervision: Progress and Achievements

In China, numerous food safety incidents happened in the past few years but, on the other hand, in this period China's food safety supervision work has also made great progress. In 2010 and 2011, both the *Special Operation against Quality Safety Problems of Agricultural Products* led by the Ministry of Agriculture and the *Special Operation Combating Illegal Food Additives* carried out by the State Administration of Food and Drug Safety have both achieved positive results. China's capability of safeguarding food safety has been significantly enhanced, and the level of food quality safety has been continuously improved. Take agricultural products for example, the statistics released by the Ministry of Agriculture show the following: In 2010, in the routine monitoring of the quality safety of vegetables, animal products, and aquatic products, the pass rates were 96.8%, 99.6% and 96.7%, respectively, maintaining a steady increase since 2009 and being over 96% for two consecutive years;

in the second quarter of 2011, in the routine monitoring of the quality safety of vegetables, animal products, and aquatic products, the pass rates were 97.9%, 99.7%, and 96.6%, respectively. This shows that China's agricultural products quality safety is overall in good condition. In addition, with respect to the food safety supervision system, China has made considerable progress in the formulation of laws and regulations, in the construction of a standards system, in quality safety monitoring and early warning of agricultural products, and in the construction of "Three Products, One Indication (pollution-free agricultural products, green agricultural products, organic agricultural products, and agro-product geographical indication)".

1.3.1 Construction of Food Safety Law System

China's food safety law system has been gradually developed since the foundation of P.R. China, and up to now it is a combination of a number of laws, like the *Food Safety Law, Product Quality Law, Agricultural Law, Law on Agricultural Products Quality Safety, Standardization Law, Import and Export Commodity Inspection Law*, and *Consumer Rights Protection Law*, and a series of complementary rules and regulations on food safety released by the State Council and ministries as well as provincial and municipal governments. Table 1.1 lists a number of laws on food safety promulgated since the foundation of P.R. China, and there are also numerous rules, regulations, and ordinances on food safety released by the State Council, ministries, and local governments at all levels. In particular, during the "Fifteenth Five-year Plan" period, China has promulgated over 70 laws and regulations relating to food safety. The issuance of the *Food Safety Law* and *Enforcement Regulations of Food Safety Law* in 2009 marks a new stage of China's food safety law system construction.

Table 1.1 Food safety laws released since the foundation of P.R. China

Date of issuance	Issuing authority	Laws	Remarks
December 2, 1986	NPC Standing Committee	Law of People's Republic of China on Frontier Health and Quarantine	Effective as of May 1, 1987
December 29, 1988	NPC Standing Committee	Standardization Law of People's Republic of China	Effective as of April 1, 1989
February 21, 1989	NPC Standing Committee	Import and Export Commodity Inspection Law of the People's Republic of China	Amended on April 28, 2002
October 30, 1991	NPC Standing Committee	Law of People's Republic of China on the Entry and Exit Animal and Plant Quarantine	Effective as of April 1, 1992
February 22, 1993	NPC Standing Committee	Product Quality Law of People's Republic of China	Amended on July 8, 2000
July 2, 1993	NPC Standing Committee	Agricultural Law of People's Republic of China	Amended on December 28, 2002
October 3, 1993	NPC Standing Committee	Consumer Rights Protection Law of People's Republic of China	Effective as of January 1, 1994
October 30, 1995	NPC Standing Committee	Food Hygiene Law of People's Republic of China	Effective as of the date of issuance
July 3, 1997	NPC Standing Committee	Animal Quarantine Law of People's Republic of China	Effective as of January 1, 1998
April 29, 2006	NPC Standing Committee	Law of People's Republic of China on Agricultural Product Quality Safety	Effective as of November 1, 2006
February 28, 2009	NPC Standing Committee	Food Safety Law of People's Republic of China	Effective as of June 1, 2009

1.3.2 Constant Improvement of the Food Safety Standard System

While China's food safety law system has been constantly improving, the construction of China's food safety standard system has been gradually strengthened.

Food safety standards cover a large number of items, including the agricultural production environment, quality of irrigation water, guidelines for the rational use of agricultural inputs, procedures of animal and plant quarantine, appropriate agricultural practices, limits of pesticides, veterinary drugs, pollutants and harmful microorganisms in food, food additives and related application standards, hygiene standards of food packaging materials, standards of special dietary food, labeling standards of food label, management and control standards of the safety food production process, and standards of food testing. These standards are involved with grain, oilseeds, fruits and vegetables and related products, milk and dairy products, meat and eggs and poultry products, aquatic products, potable spirit, condiments, infant foods, and other edible agricultural products and processed foods, covering all stages from food production, processing, and distribution to the final consumption. At present, China has released more than 1,800 national standards concerning food safety and more than 2,900 industry standards, including 634 mandatory national standards.

1.3.3 Capacity Enhancement of Food Safety Monitoring and Early Warning

Agricultural products are the basis of the food industry and are also important sources of food safety risks. During the "Eleventh Five-Year" period, China implemented the *Construction Plan of National Agricultural Product Quality Safety Inspection System (2006-2010)* and on the whole established a quality safety monitoring network, which covers the major cities across the country, the main agricultural production areas, and the main varieties of agricultural products. *The Twelfth Five-Year Plan for the Development of Agricultural Product Quality safety* developed by the Ministry of Agriculture shows that: during the "Eleventh Five-Year" period, the government invested RMB 5.9

billion Yuan for the platform construction of agricultural products – newly constructed, renovated and expanded 49 ministerial-level quality inspection centers, 30 provincial-level comprehensive quality inspection centers, 936 county-level agricultural products quality inspection stations; the range of routine monitoring on consumption safety has covered 138 large and medium-sized cities across the country, 101 kinds of agricultural products, and 86 safety test parameters.

1.3.4 Steady Advancement of the Project of High Quality Safety Brand Building: Three Products, One Indication

"Three Products, One Indication" project is involved with pollution-free agricultural products, green agricultural products, organic agricultural products, and agro-product geographical indication. As the focus of a national high quality safety brand building project and as the characteristic content of the construction of agricultural modernization, the total scale of "Three Products, One Indication" project is steadily increasing. By the end of 2010, up to 56, 532 pollution-free agricultural products, 16,748 green agricultural products, and 5,598 organic agricultural products were certified and 535 agro-product geographical indications were newly registered. The amount of production area certified with "Three Products, One Indication" accounts for more than 30% of the total amount of agricultural production areas; the amount of certified products account for over 30% of the total amount of edible agricultural products. In the first five months of 2011, the "Three Products, One Indication" project maintains the momentum of rapid development: 9,645 agricultural products of 4,894 production units are certified with pollution-free agricultural products certification, with an annual output of 33.8176 million tons; 601 companies and 1,299 products are newly conferred with the right to use the mark of green agricultural products; 365 enterprises and 1,846 agricultural products have gone through organic food certification; 144 agro-product geographical indications have passed expert review and public notice.

On the whole, as the Communist Party and the country in recent years attached great importance to food safety issues, China's security system for food safety has been constantly improved, the capacity of ensuring food safety has been greatly enhanced, the overall level of food safety has been raised, and the level of consumers' food safety information has been gradually restored and increased. However, apart from the affirmation of the current work, we should clearly realize the fact that China's food safety supervision work is extremely arduous and the supervision system is still not sound enough. This requires that we should accurately analyze the causes of China's food safety problems, primarily grasp the key control points of the industrial chain, efficiently allocate supervision resources, and fully draw on international experience to further improve the construction of China's food safety supervision system.

1.4 China's Food Safety Supervision: Problem Analysis

In 2009, Food Safety Survey (2009) conducted by Hexun Net showed that about half (48.7%) of the 613 interviewees in Beijing, Shanghai and Guangzhou expressed great concern about food safety, and only 1.3% of them expressed no concern; nearly 40% (39.9%) of the interviewees thought that the food safety situation was getting worse, while those who believed that the food safety situation would get better accounted for only 24.7%. Although this data reflecting the public judgment for the food safety situation may be influenced by the frequent outbreak of a series of incidents, like the 2008 Melamine Incident, it is not difficult to see that the endless stream of food safety incidents has negative impacts on public confidence in food safety. According to the data from the National Center for Food Information and National Food Safety Resource Database as well as the data from related media coverage, Liu *et al.* (2011) found that during the decade from January 2001 to January 2010, up to 1,460 food quality safety incidents occurred. According to the statistics of ZCCW Net[2], from 2004 to 2011, up to 17,268 news articles on food safety incidents can be found, and among which up to 2,849 articles

---

[2] ZCCW Net (http://www.zccw.info/) is news database created by Wu Heng, a graduate student of Fudan University, for the purpose of helping the public to know more about China's food safety situation.

clearly showed the place of incidents, the type of food involved, the reasons why such food is harmful to human beings, and other key information. Although, in recent years, China's food safety supervision system has continued to improve and the overall level of food safety situation has improved, yet a total of more than a thousand successive food safety incidents, like Fuyang Tainted Milk Incident, Melamine Incident, the Lean Meat Powder Incident, and the Beef Extract Incident, indicate that the root of China's food quality safety incidents is complicated and it is difficult to govern such food safety issues.

Why do food safety incidents happen one after another? Viewed from the surface phenomena of such incidents, the influencing factors of food safety incidents can be broadly divided into the following categories: (1) pollution of environmental resources which is the root of agricultural production, such as agricultural non-point source pollution; (2) residues of chemical fertilizers, pesticides, growth hormone, and other harmful chemicals which are used in the production process of crop farming, fish breeding and poultry raising, such as the Poisonous Cowpea Incident, Turbot Fish Incident, and other incidents of pesticide residues in vegetables; (3) illegal or excessive use of food additives in the processing and storage of agricultural products, such as the representative Melamine Incident, and the majority of food safety incidents are derived from this; (4) microbial contamination in the production process of agricultural products, such as the representative Golden Apple Snail Incident and other food-borne illness; (5) food safety risks resulting from new materials and new techniques, such as the current popular concern on the safety issues of genetically modified food. Apart from these characterization factors, it is not difficult for us to realize that the problems of food safety standards, management factors, and even institutional factors are the deeper reasons for the current frequent food safety problems. On the whole, we believe that the reasons for China's food safety problems can be attributed to three interrelated aspects: the subjective and intentional criminal acts taken by the behavioral agents of food supply chain for personal gain; imperfect food safety supervision system and inadequate supervision; technical factors, primarily finding expression in their limitations to the monitoring of food safety as well as in the new quality safety risks arising from the use of new technology in food production. Therefore, the following part of this chapter analyzes the real causes of food safety issues from three aspects, namely, the food supply chain, the national food safety supervision, and the influence of new technology on food safety.

1.4.1 China's Food Supply Chain

1.4.1.1 *Food Supply Chain and Food Safety*

The concept of the food supply chain, which was first put forward by the American scholar Zuubier on the basis of a general supply chain, refers to a vertical integration mode of operation, through which the production and sales organizations of agricultural products and food commodities reduce logistics costs, improve the quality of agricultural products and food, and enhance their service level. The food supply chain can, on the one hand, effectively meet the current consumer demand for fresh and safe food, especially agricultural products, and on the other hand it fits into the requirements of food quality safety laws and regulations; therefore, it evolves rapidly in the developed world. Meanwhile, as the food supply chain covers all aspects of food production, processing, storage, transportation, distribution, sales, and consumption, and as it is involved with producers, distributors, consumers, and other behavioral agents, the control and supervision of food safety throughout the food supply chain has become the research focus of the world.

The methods of food safety supervision and, in particular, the tracking of food information are closely linked to the types of the food supply chain. Based on the length of the supply chain, the distance between the place of production and the market, the number of behavioral agents at both ends of the supply chain, and other characteristics, foreign scholars usually divide the food supply chain into four categories: dumbbell-shaped food supply chain, T-shaped food supply chain, symmetric food supply chain, and mixed type food supply chain. For food supply chains which are involved with loose cooperation or shorter chain length, such as dumbbell-shaped and T-shaped food supply chains, the information chain should be used to track the food information; however, for food supply chains which are involved with closer cooperation or longer chain length, such as symmetric and mixed type food supply chains, the value chain should be used to track the food information. In China, although the four types of food supply chains exist simultaneously, yet dumbbell-shaped and T-shaped food supply chains dominate the mainstream positions and the T-shaped food supply chain is widely used in the agricultural products logistics system which includes the wholesale market as the core.

No matter what type of food supply chain it is, it is related to many behavioral agents and a variety of stages. Therefore, the behavior of any participant may affect the food quality on the food supply chain. In addition, the characteristics of food commodities, especially agricultural products, on the food supply chain, such as large volume, low value, perishability, and short consumption cycle, lead to more stringent requirements in the processing, freezing, storage and transportation of food commodities and especially agricultural products on the food supply chain, and to the smoothness of all stages of the food supply chain compared to general industrial products. However, in China, the current food production and processing is scattered and of small-scale, the traffic infrastructure is underdeveloped, the port refrigeration equipment and refrigerated storage facilities are inadequate, management methods lag behind, and modern IT platforms for the supply chain have not been established; these problems greatly restrict the quality and efficiency of the operation of food supply chains in China and also, to some extent, exacerbate China's food safety incidents. In short, in order to further improve the food safety supervision system based on the food supply chain, we should sort out the relationship between different behavioral agents at different stages of the supply chain, analyze the causes of food safety incidents at each stage, identify the critical control points of the food supply chain management, and then carry out targeted control and supervision.

1.4.1.2 *Influencing Factors of Food Safety Issues at Each Stage of the Food Supply Chain and Cause Analysis*

In order to facilitate the analysis, we first establish a "farm to fork" typical food supply chain, as shown in Fig. 1.1. This complete supply chain covers all stages of food production, processing, storage, transportation, distribution, and consumption, and is involved with food producers, processors, wholesalers, retailers, consumers, and other behavioral agents. Then, in accordance with this typical food supply chain, we specifically analyze the influencing factors of food safety issues at each stage, and then identify the key control points of food supply chain management.

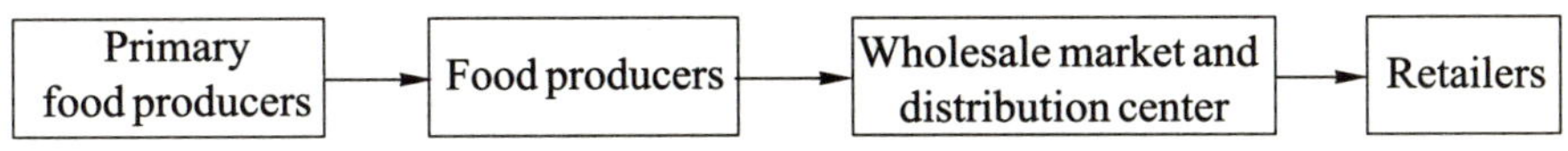

Fig. 1.1 A typical food supply chain

Production Stage

Agricultural products are the main source of the food industry, and agricultural production is the basis of the food supply chain. Viewed from the characterization reasons of food safety issues, the problems at this stage mainly result from the polluted environment necessary for agricultural production and from the excessive use of chemical fertilizers and pesticides in the production process of crop farming as well as the use of illegal feeding-stuff in fish breeding and poultry raising. According to the China Statistical Yearbook (2011), statistics show that, from 1978 to 2010, the effective irrigated area increased from 44.965 million ha to 60.3477 million ha, an increase of 34.2%; at the same time, the fertilizer application rate increased from 8.84 million tons in 1978 to 55.617 million tons in 2010, an increase of 529.2%. According to calculations by experts, currently in China the average fertilizer application rate reached 434.3 kg/ha which is 1.93 times the internationally recognized safe rate of fertilizer application 225 kg/ha. But the deeper reasons lie in the inadequacy or even nonexistence of food safety supervision at the production stage of food. Specifically, firstly due to the long-term urban-rural dual structure, the very limited capacity and resources of food safety supervision are mainly allocated in urban areas, while in rural areas the range of food safety supervision is narrow and the effort of food safety supervision is inadequate. Secondly, the food safety supervision work is involved with multiple departments, while the division of responsibilities among the various departments is not clear, so that they do not work in a coordinated manner. This, on the one hand, causes repetitive allocation and inefficient use of supervision resources, and on the other hand, results in the situation that all departments scramble for profitable matters and shrink away from unprofitable matters; accordingly, hidden perils of food safety cannot be dealt with in a timely manner. Last but not least, from an objective point of view, currently China's behavioral agents of agricultural production show a series of characteristics in that the organization level of such agents is low, there are a large number of such agents but they are small in size and scattered, and their ability to violate laws and regulations are hidden. These characteristics cause great difficulties in the supervision of the production process.

Processing Stage

Viewed from the statistics on currently exposed food safety incidents, most such incidents occurred in the processing stage. In terms of characterization factors, the problems are mainly the use of chemical preservatives in the early processing stage, the use of illegal additives in the deep-processing stage, the microbial contamination caused by a dirty processing environment, and the hidden safety hazards due to the use of new production technology. In 2011, in the **Special Operation against "Lean Meat Powder"** led by the Ministry of Agriculture, only in the first phase of five months, 120 cases of illegal manufacture, sale, and use of lean mean powder were investigated, and more than 980 criminal suspects were arrested. However, on the one hand, the problems at this stage are because of the subjective and intentional criminal acts of producers and, on the other hand, the supervision system can hardly absolve itself from the blame. Analyzed from the perspective of the management system and institution level, the problems mainly concentrate on the following aspects: firstly, because the primary stage of food processing has little value added and the circulation speed is relatively fast, the primary stage is often ignored by administration authorities. Secondly, the level of food quality safety in the processing stage is often subject to the dual constraints of the level of food safety in the production process and of the quality requirements of the marketing stage, which makes food processing enterprises reduce their production costs; in addition, the inadequate external supervision also provides an ongoing opportunity for processing enterprises to use illegal chemical additives and reduce their processing sanitation level. Thirdly, the deep processing stage of food is involved with a lot of manufacturing processes and procedures, so it is more difficult to control. Finally, the food industry does not need too much capital and high technology and the barrier to entry is relatively low, resulting in the fact that there are a large number of food processing enterprises in China and there are quite a lot of small businesses and small workshops at different levels; thus, China's food safety supervision is very difficult. The *Twelfth Five-Year Plan for the Processing Industry of Agricultural Products* shows that the above-scale enterprises account for only 24% of all agricultural products processing enterprises, and among which the annual income of 21 enterprises exceeds RMB 10 billion Yuan, that of 4 enterprises exceeds RMB 50 billion Yuan, and that of only 1 enterprise exceeds RMB 100 billion Yuan. According to the Chinese Food Industry Analysis Report (2011) released by China Economic Net, from January to March 2011, China's food manufacturing industry realized a total sales income of RMB 285.507 billion Yuan. Large enterprises realized a total sales income of RMB 51.495 billion Yuan, accounting for 18.04 percent of that of the country; medium-sized enterprises realized a total sales income of RMB 112.872 billion Yuan, accounting for 39.53 percent of that of the country; small enterprises realized a total sales income of RMB 121.14 billion Yuan, accounting for 42.43 percent of that of the country. That is to say that although in recent years the state has increased its efforts to govern food safety issues of small and medium-sized enterprises and the scale of the food industry has been greatly enhanced, yet the sales income of small and medium-sized food processing enterprises still accounts for a large percentage of the aggregate sales income of the food industry.

Storage and Transportation Stage

Viewed from the characterization factors, food safety incidents occurring in the storage and transportation stages are more than the microbial contamination incidents caused by dirty storage and transportation environment as well as the food spoilage and deterioration incidents caused by the poor circulation of the supply chain. The emergence of these incidents is related to the currently imperfect food logistics system. Firstly, a convenient transportation network that has reasonable layout and complete functions has not been established, which directly restricts the operational efficiency of the food supply chain in China, and which not only increases the circulation costs, but also reduces the circulation speed of food commodities. Secondly, the refrigeration equipment and refrigerated storage facilities within China's food supply chain are grossly inadequate. The data of *Development Plan for Cold Chain Logistics of Agricultural Products* show that, currently, there are only 20,000 refrigerated storage units in China with a cold storage capacity of only 8.8 million tons, while in China there are about 400 million tons of fresh agricultural products entering into circulation each year. At present, the cold chain circulation rate of vegetables, meat, and aquatic products is 5%, 15%, and 23%, respectively, and the refrigerated transport rate is 15%, 30%, and 40%, respectively. Compared with developed countries in Europe and the United States, the gap is more obvious: in Europe and the United States and other developed countries, the long-distance cold chain transport rate of perishable agricultural products is 100%, while that of China is only 32%; in Europe and the United States, the decay rate of rail transport is controlled between 5% – 8%, while that of China is up to 25%—30%. This consequently leads to higher food wastage and safety risks in the storage and transportation stage. Last but not least, because food logistics has higher requirements for capital and technology, therefore, apart from a few big food companies that have their own storage equipment, the vast majority of food suppliers depend on third-party logistics companies. In addition, in our country, apart from the imperfection of the logistics infrastructure, the large number of logistics companies may also lead to frequent handover of food commodities in the storage and transportation stage, which will inevitably cost much time and cause much food wastage.

Wholesale Sales Stage

The food safety hazards in this stage are mainly due to the poor infrastructure of the wholesale market and especially the lagging behind of cold chain construction, poor sanitary conditions, and weak enforcement of food safety supervision. In the food circulation process, the wholesale market is one of the main channels, and it is also the core stage of the convergence of product flow and information flow on the food supply chain. Therefore, it is an inevitable choice to carry out food safety supervision and establish a tracking system of food quality safety information with the wholesale market as a key control point. However, the majority of food wholesale markets, especially the agricultural products wholesale market, are in excessive pursuit of economic efficiency, but disregard social benefits: for example, such wholesale markets pay much attention to the site area and customer flow volume, pursue the degree of popularity, emphasize the market distribution of goods, price formation, supply and demand adjustment, settlement, and other functions that are directly linked to their own economic interests, but ignore safety supervision that may increase their operating costs. Of course, this is also the result of the government's inaccurate function positioning and the private mode of operation of food wholesale markets.

Consumption Stage

Food safety incidents rarely occur in this stage, and the few problems may mainly arise from an unscientific cooking process, irrational diet, and/or unsanitary eating environment.

Integrating the analysis of influencing factors at each stage of the food supply chain and together with the distribution of 1,460 food quality safety incidents at all stages analyzed by Liu *et al.* (2011) and which happened during January 2001 to January 2010 (Table 1.2), we can clearly see that the food safety incidents occur mainly in the food production stage and the food processing stage, which should become the key control points of food safety supervision in the supply chain. As the core stage of the convergence of product flow and information flow in the food supply chain, the food wholesale market greatly influences the food safety supervision with its food safety information distribution and security control and, therefore, it should also be considered as a key control point of food safety supervision.

Table 1.2 Distribution of 1,460 food quality safety incidents happening during January 2001 to January 2010

Stage	Amount (incident)
Production of agriculture products	199
Primary processing of agriculture products	193
Deep processing of agriculture products	1296
Storage and transportation/ marketing/catering	209
Consumption	6
Total	1903

Source: Modified from Liu et al.(2011) (The total is 1,903 that is greater than 1,460 the effective number of incidents, for part of the incidents reflect the problems of many stages).

1.4.2 Traceability of China's Food Quality Safety

The food quality safety problem in nature is the market distortion due to asymmetric information (Caswell, 1998; Ritson and Mai, 1998). Therefore, an important means to overcome the market failure is to increase the effective supply of food safety information, so as to encourage the producers and operators to raise their level of control over food safety and quality by means of the "better quality, higher price" incentive mechanism or the potential punishment mechanism on the basis of clear responsibility. The frequent outbreaks of food-borne serious incidents, like mad cow disease in Europe, triggered a crisis of consumer confidence and a crisis of confidence in the government. This research conclusion was first adopted by the EU and other developed countries. Through legislation, developing mandatory standards, and resorting to a market access system they, one after another, have established a farm-to-fork food safety traceability system, which is regarded as an important part of the food safety supervision system. In the event of food safety incidents, through the food safety traceability system, they can quickly and accurately position the food that has the problem and the corresponding stage of the supply chain, and thus they can effectively implement a food recall to avoid the proliferation of food safety incidents.

After years of development of the food safety traceability system, developed countries have established relatively complete and specialized laws and regulations, quality standards, and a network technology system. However, in China, the construction of a food safety traceability system has just begun. In 2000, China formally began to establish a traceability management system, and ensuring food safety is regarded as the focus of the traceability system for food safety supervision. On May 24, 2002, the Ministry of Agriculture issued *Management Regulations for Animal Vaccination Identification Tag* (Order No. 13), which prescribed that pigs, cows, and sheep must wear vaccination identification ear tags, and a relevant management system of vaccination records should be established. In 2003, the State General Administration of Quality Supervision, Inspection and Quarantine implemented "China Barcode Propulsion Engineering". In 2004, the Ministry of Agriculture launched the "Beijing Vegetable Product Quality Traceability System". In 2006, *Law of the People's Republic of China on Agricultural Product Quality Safety* was officially launched to control and supervise the whole process of agricultural production for the sake of agricultural product quality safety. Such control and supervision work over agricultural product quality safety has been implemented in eight pilot provinces and cities, and the key point is to establish agricultural production records. In August 2007, China officially released and began to implement *Management Regulations for Food Recall*. Since November 29, 2007, the State General Administration of Quality Supervision, Inspection and Quarantine, the Ministry of Commerce, and the State Administration for Industry and Commerce have resorted to mandatory requirements in nine categories covering 69 kinds of major products in the country, like food, household appliances, and cosmetics; such products must have an electronic quality supervision code, or otherwise they are not allowed to enter into the market. Local places, such as Beijing, Shanghai, and Nanjing, have also carried out some pilot work of constructing a quality safety traceability system. After years of development, China has achieved encouraging progress in the construction of a food quality safety traceability system: First, supporting laws and regulations are being continuously improved. Second, a series of related quality safety standards have been promulgated, and barcode food safety tracing platforms have been established. With respect to traceable food and enterprises, a number of traceability subsystems have been established on these platforms. Wholesale marketing of agricultural products is the core stage of food circulation and especially the circulation of agricultural products in China, and therefore, the construction of a traceability system in the agricultural products wholesale market largely reflects the status of constructing a food safety traceability system in China. *Circulation Stage Food Safety Investigation Report (2008)*, which is issued by the Ministry of Commerce, shows that only 63.6 percent of the wholesale market of the 1,919 urban markets and 1,835 rural markets surveyed have an electronic standing book, and 28.1 percent of them have realized a unified settlement. In addition, Zhou and Zhang (2011) found that, with respect to the construction of a vegetable quality safety traceability system, the quality safety of only 46.6% of vegetables in the production stage is traceable, only 45.7% in the circulation stage, and only 33.5% in the consumption stage. Although most of the vegetable wholesale market has established a sound supporting system for a traceability system, in order to avoid suppliers and buyers moving to other wholesale markets due to cost increase caused by quality safety tracing, currently wholesale markets are reluctant to take the lead in the implementation of a quality safety traceability system, especially those at county level and below.

However, there are still great difficulties to overcome in order to further improve a food quality safety traceability system. To establish a food quality safety traceability system, China is facing six major obstacles: first, the production of food and especially agricultural products is scattered, the production intensity is not high, and the level of standardization and technology is low; second, the method of food distribution is still relatively backward, traditional distribution channels, such as wholesale markets and bazaars, still occupy a considerable proportion, and modern distribution channels, such as chain supermarkets, are still not widely available; third, a food safety law system and standard system are not perfect, relevant regulations and standards are deficient or lag behind the real needs, and a lot of standards or regulations are not in line with international practice; fourth, the food safety supervision system is not systematic and unified enough; fifth, the public as a whole has not fully realized the importance of a food safety traceability system; sixth, the cost of establishing the traceability system is relatively high, and most enterprises are lacking the momentum for initial investment. In addition, the emerging food safety incidents in recent years have also revealed that there are also some other problems in the construction process of China's food quality safety traceability system: first, China's food quality safety traceability system mainly adopts separate database and information inquiry platforms, and the information of different databases does not follow unified standards and thus is lacking a universal property, which hinders the establishment of a food safety traceability system that can cover the whole process of the food supply chain; second, the relevant laws and regulations are still not perfect, the standard system is not sound enough, the enforcement of food safety laws and regulations is lax, and corresponding punishment does not have enough of a deterrent force; third, the management is in chaos and the responsibility is unclear. Currently, China mainly uses a segment management mode and multiple departments function in parallel, but the functions of different departments are not clearly defined and there is no specialized institution to coordinate the work of various departments. Zhou and Zhang (2011) studied and found that the government is the main force to promote the construction of a traceability system, but it cannot coordinate the various markets to act simultaneously; the government can only directly control state-owned or collective ownership wholesale markets, and it is lacking a strong control over private wholesale markets. Currently, in China, the construction of an agricultural products quality safety traceability system, the core of which is the wholesale market, lacks the participation of all behavioral agents within the supply chain of agricultural products. In the upper reaches of the wholesale market, the farmers lack the awareness of registering agricultural products quality safety information and the awareness of providing vouchers or invoices; in the lower reaches of the wholesale market, the management of the farm product market is lax, and consumers' awareness of asking for invoices still needs to be enhanced. All in all, there is still a lot of work to do to establish a perfect food safety traceability system based on the food supply chain.

1.4.3 China's Food Quality Safety Supervision

The public goods attribution of food safety issues determines that the market often fails to ensure food safety. Therefore, the government should establish a perfect food safety supervision system through promulgating laws and regulations, building quality safety standard systems, allocating regulatory agencies, personnel and funds, and promoting the implementation of this supervision system through optimizing the management system on the basis of perfecting the supervision system. Meanwhile, the government should also effectively combine market means with administrative measures, stimulate and enhance the enthusiasm of all behavioral agents within the food supply chain for improving product quality safety through market means and security measures, and ultimately build a complete government-led food quality safety supervision system which is subject to the full participation of third parties.

1.4.3.1 *China's Food Safety Laws and Regulations System*

China's food safety legislative work has made considerable progress. With respect to the serious conflict between existing food safety laws and regulations, the inefficiency of the supervision system, the disunity of safety standards, and lenient punishment, the *Food Safety Law of People's Republic of China* released in 2009 has made targeted adjustments: first, the "multiple-start segmented supervision mode", which has been much criticized for a long period of time, is adjusted to segmented supervision mode under the unified leadership of the People's Government at all levels; in accordance with the segmented supervision mode, the supervision departments at all levels supervise all stages of the food supply chain; and the responsibilities of all departments are being gradually clearly defined. Second, it has been clarified that food producers are the primary behavioral agents responsible for food safety, their legal obligations are clearly defined, the cost of their unlawful behavior has been raised, and thus food safety can be guaranteed from its source. Third, a food recall system has been determined. Fourth, the responsibilities of supervisors are clearly defined, as to in which cases such supervisors fail to fulfill their duties on food safety supervision and when their behavior results in food safety incidents; the due responsibilities of any third parties are also clarified, if such third parties as inspection and quarantine authorities, news media, stars and celebrities, and other social organizations violate laws and regulations. But compared with a mature foreign food safety supervision system, there is still a big gap. The gap is mainly reflected in the spirit of legislation, the determination of responsible behavioral agents, the guidance and standardization of the behavior of the supervision body, as well as risk prevention. To sum up, China's food safety laws and regulations system has a series of problems, such as biased legislative value, supervision failure that is difficult to effectively solve, inadequate guidance to the behavior of the news media and industry associations, failure to effectively guide consumers to play supervisory functions, an incomplete punishment system to address the bad food safety reputation, a limited number of food quality standards, and food quality standards lacking universal properties.

1.4.3.2 *China's Food Safety Supervision Mode and Supervision Body*

The food safety supervision body refers to institutions that can exercise the power of food safety supervision. Since 2003, China's food safety supervision system has experienced four types of adjustments, but it still continues with the "segmented supervision first, variety supervision second" multi-department supervision principle and multiple-start segmented supervision mode, i.e. different supervision bodies are responsible for the supervision work at different stages of the food supply chain. Although *Food Safety Law* has adjusted the "multiple-start segmented supervision mode" to a segmented supervision mode under the unified leadership of the People's Government at all levels and in accordance with the segmented supervision mode, the supervision departments at all levels supervise all stages of the food supply chain, i.e. a change from segment supervision to variety supervision, yet this kind of decentralized management mode has not been changed. Although, to a certain extent, the decentralized supervision mode can meet the requirements of China's current situations, such as the long food supply chain, too many chain stages, and complex circumstances, this kind of decentralized mode will inevitably predicate the multiplicity of supervision bodies. According to relevant laws and regulations for food safety supervision, China's food safety supervision body is involved with health administration departments, food and drug administration, agricultural departments, commerce departments, public security departments, industry and commerce departments, quality supervision departments, customs departments, entry-exit inspection and quarantine departments, marine fishery departments, economic and trade departments, and environmental protection departments, etc. If the duty of these many supervision bodies cannot be clearly defined, and if the work of these departments cannot be effectively coordinated, the problems of the original segmented supervision mode still cannot be avoided, including high supervision cost, supervision confusion, supervision corruption, buck-passing, and the absence of a rural supervision body. Thus, regardless of what kind of food safety supervision mode is established, it is imperative to clearly define the scope of responsibility of the supervision bodies, straighten out the relationship between supervision bodies, and determine the accountability system of supervision bodies, so as to ensure the effective functioning of the supervision system. In addition to traditional supervision of government departments, the role of industry associations, media and other social third-party agencies should be given full play. The absence of a supervision body for rural food safety is rooted in the uneven allocation of resources under long-term urban-rural dual structure and the resulting serious shortage of supervision personnel and funds. Coupled with the low level of consumption of rural residents, inadequate awareness of food safety problems, and poor access to information, the rural areas tend to become the blind areas of food safety supervision.

1.4.3.3 *China's Unsafe Food Recall System*

Unsafe food recall refers to the practice that defected food that does not comply with food safety standards and may endanger public health and safety is recalled from sales, distribution and the consumption fields. The unsafe food recall system is an important part of China's food safety system, and its degree of perfection is directly dependent on the level of recall legislation and the level of law enforcement. Currently, China's unsafe food recall system is built and implemented in accordance with the *Food Safety Law, Regulations on the Implementation of Food Safety Law,* and *Management Regulations of Food Recall.* Due to the lack of an integrity mechanism, currently in China the effect of food recall is bound to be greatly discounted if totally dependent on the voluntary recall of companies. Therefore China's current recall system is a combination of voluntary recall and mandatory recall. Although, in the *Management Regulations of Food Recall,* there are clear provisions for the recall conditions, recall method, recall class, the content of the recall program, and the obligations of behavioral agents for food recall, yet there are still many problems in China's food recall system: first, the inconsistencies existing in *Food Safety Law* and *Management Regulations of Food Recall* may easily lead to confusion of law enforcement. For example, Article 25 of *Management Regulations of Food Recall* prescribes that the behavioral agent that orders the recall is the "State General Administration of Quality Supervision, Inspection and Quarantine", but Article 53 of *Food Safety Law* prescribes that "In the event that a food producer or trader fails to recall or stop trading in the food that does not comply with the food safety standards as required in the Article, the executive departments of quality supervision, industry and commerce, and food and drug administration at the county level or above may order it to recall or stop trading in the food". Second, the duties of the supervision body are not clearly prescribed. For example, there are no detailed rules on how to assess and how to determine the recall, which may lead to too large a discretion of the executive authorities and unlimited penetration of government accountability in the private sector. Third, the effective implementation of the recall system depends on the supporting food safety information collection system, the food quality safety traceability system, and effective law enforcement, but there are also problems in these aspects in China.

1.4.4 The Application of New Technology and Food Safety: Take Transgenic Technology as an Example

In the analysis of the role of technology in production, traditional economic theory often focuses on analyzing economic benefits, i.e. its impact on the yield increase in production, so we usually regard technological progress as the most important factor in agricultural development. However, as consumers increasingly attach importance to their own health and the improvement of their living environment, the evaluation criteria of new technology has not been merely limited to the economic benefits of the new technology, but rather the ecological and social benefits of new technology have also been paid more and more attention. In the field of food safety, people's attitudes to the application of transgenic technology in agricultural production fully demonstrate the changes in new technology's evaluation criteria.

Transgenic technology is to introduce artificially separated and modified genes into the genome of organisms, so as to achieve the purpose of modifying organisms. If transgenic technology is applied to agricultural production, there will be genetically modified (GM) foods. There is no doubt that the GM agricultural products have the advantages of biological characteristics that traditional agricultural products cannot compete with: rapid growth, disease-resistance, pest-resistance, herbicide-resistance, anti-bad weather, more delicious, and more nutritious. This means that GM crops will not only help to improve the yield and quality of agricultural products and reduce production costs and labor intensity, but also can greatly reduce the use of pesticides in agricultural production and thus alleviate the problem of polluting the agricultural environment. It is because of these advantages that GM foods begin to gain more popularity in the world. In 1996, when GM crops were first promoted for commercial planting, there were only six countries that wanted to plant such crops, with the total plant area of about 1.7 million hectares. But in 2009, 14.4 million farmers in 25 countries (including 16 developing and 9 developed countries) planted 134 million hectares of 10 varieties of GM crops, like soybeans, corn, cotton, and rapeseed; in the whole world, the plant area of GM crops reached nearly 1 billion hectares, an increase of nearly 80 times over 1996. According to statistics, in 2009, the GM variety of soybean accounted for 77% of the total plant area of soybean, 26% of corn and 21% of rapeseed.

In sharp contrast to the rapid development of GM crops, consumers in most countries are cautious about GM food. Although previous studies show that the United States and India, a representative of some developing countries, are more likely to accept GM foods, yet a study made by the International Food Information Council shows that although GM crops have been marketed for 10 years in the United States, there is still a lot of confusion and bias in the understanding of GM crops among consumers. The European Union (EU) has always strictly controlled GM food, and the EU has prescribed strict rules for the potential risks, commercialization, and environmental release of GM food. The consumer acceptance of GM foods in the EU is generally low. The reasons why consumers hold such an attitude to GM food, on the one hand, lies in social ethics and religious beliefs, but on the other hand, I think, a more common reason is that people tend to keep away from the apparent risks caused by GM foods to human health, agricultural production environment, and the ecological environment. In other words, people attach more and more importance to food safety and nutrition, not just the quantity of food, which is also the major feature of current food consumption. Chinese consumers should rationally and comprehensively measure the role of transgenic technology in agricultural production and gain a clear understanding of the food safety potential risks.

1.5 Conclusions

In recent years, China has taken a number of effective measures to strengthen the supervision of food quality and safety, but food safety incidents still occur sometimes. The recurrence and intractability of such incidents suggest that, in addition to the imperfect supervision system, the greatest obstacle to China's food quality safety management is that China's "farm to fork" food supply chain has too many stages, the members in the supply chain have not formed a stable, strategic and cooperative relation, and on the other hand, during the transitional period, some practitioners lack social responsibility. Therefore, China's food quality safety management and the establishment of food quality and safety traceability system cannot directly adopt the current experience of developed countries; they should, on the one hand, follow the development trend of international food quality and safety supervision, and should, on the other hand, combine with the establishment of China's agricultural industrialization and standardization, integrate China's existing but isolated effective measures, such as the establishment of bases for the implementation of the system of claiming certificates or invoices, for the performance of Management Regulations for Pig Slaughtering and Quarantine Inspection in Designated Places, and for the conduct of World Expo, as well as the establishment of a market access system, taking into consideration the demand, the dynamic mechanism, and the performance of important measures of food supply chain members for food quality and safety control, as well as the difficulties and the deep-seated reasons in the implementation process of such measures.

To this end, this book chooses important agricultural products like vegetables, pork and aquatic products as the subjects to be investigated. From an "integrated" vertical perspective of the supply chain and according to the degree of industrialization of different products, focusing on the key links of quality and safety control of vegetables, pork and aquatic products, this book carries out empirical analysis of the construction of the food quality and safety control system, such as the HACCP (Hazard Analysis Critical Control Point) quality control system and the food quality and safety traceability system, deeply analyzes and straightens out the dynamic mechanism and the performance of different business entities implementing the food quality and safety management system, as well as the bottleneck and deep-seated causes of promoting advanced experiences of pilot areas and enterprises in China, and puts forward ideas and suggestions for establishing long-term effective food quality and safety management systems with regard to vegetables, pork, and aquatic products, which can provide a scientific basis for the government to design food quality and safety management policies. The rest of this book is organized as follows:

Chapter 2 presents safety of vegetables and the use of pesticides by farmers in China. Chapter 3 discusses adoption of food safety and quality standards by China's agricultural cooperatives. Chapter 4 covers Implementation of Food Safety and Quality Standards by the Vegetable Processing Industry. Chapter 5 presents a comparative analysis on adoption of a HACCP system in the Chinese food industry. Chapter 6 explores an empirical analysis of the implementation of vegetable quality and safety traceability system centering on wholesale markets. Chapter 7 discusses the investment in voluntary traceability, an analysis of the Chinese hog slaughterhouses and processors, and Chapter 8 gives a perspective on self-inspection behavior, quality perception and quality control behavior in aquaculture. The last section provides an outlook for China's food safety and situation and policy recommendations.

References

Caswell, J.A. (1998). Valuing the benefits and costs of improved food safety and nutrition. *Australian Journal of Agricultural and Resource Economics*, 42(4), 409-424.

Chinese Food Industry Analysis Report (2011). China Economic Net. http://cei.gov.cn/.

China Statistical Yearbook (2011). National Bureau of Statistics of the People's Republic of China. Beijing: China Statistics Press.

Circulation Stage Food Safety Investigation Report (2008). The Ministry of Commerce. http://scyxs.mofcom.gov.cn/aarticle/dwmyxs/i/200905/20090506237273.html.

Liu, C., Zhang, H. & An, Y.F. (2011). Study on weaknesses, root causes and key issues of China's food quality safety: Based on the empirical analysis of 1,460 food quality safety cases. *Issues in Agricultural Economy*, (1), 24-31.

Ritson, C. & Mai, L.W. (1998). The economics of food safety. *Nutrition &*

Food Science, 98(5), 253-259.

Survey on Food Safety (2009). Hexun Net. http://news.hexun.com/2009-03-05/115316289.html.

Yu, X.F., et al. (2006). *Introduction to Non-traditional Security*. Hangzhou: Zhejiang People's Publishing House.

Zhou, J.H. & Zhang, S.D. (2011). The construction of vegetable quality safety traceability system with wholesale market at the core: Based on a two-dimensional perspective of suppliers and relevant administrative departments. *Issues in Agricultural Economy*, (1), 32-38.

Safety of Vegetables and the Use of Pesticides by Farmers in China

2.1 Introduction

For the last twenty years, pesticides have been used extensively to increase crop yield and produce high quality products for consumption in China (Widawsky *et al.*, 1998). According to China Statistics Yearbook (2007), the total amount of chemical pesticides produced in China increased from 201,000 tons in 1985 to 1384,600 tons in 2006. Huang *et al.* (2003) stated that Chinese farmers apply more chemical pesticides to their crops than producers in almost any other country in the world. It has been reported that the substantive application of pesticides may cause pesticide residues in food crops, which is especially true for vegetables, which among other crops receive the highest application of pesticides (Ngowi *et al.*, 2007). Pesticide residues in vegetables not only pose problems for international trade but also damage the health of Chinese consumers.

In recent years, China's exports have suffered due to vegetable safety issues, as food safety standards in developed countries such as Europe, Japan and the U.S. are more strictly enforced (Jin *et al.*, 2008; Calvin *et al.*, 2006). For example, in 2002 Japan revised the Maximum Residual Limit (MRL) of the pesticide Chlorpyrifos in spinach from 0.1 ppm to 0.01 ppm. As a result, the export of spinach from China to Japan dropped dramatically, from a high level of US $33.89 million in 2001, to US $14.3 million in 2002 and US $3.95 million in 2003 (Chen *et al.*, 2008).

Pesticide residues in vegetables pose a risk to the health of consumers in China as Chinese people consume a huge amount of vegetables. According to Statistics of the World (2008), China, after Greece, has the second highest annual per-capita consumption of vegetables in the world. In 2003, consumption of vegetables per-capita in China reached 270.49 kg compared with the average world per-capita vegetable consumption of 94.45 kg (Statistics of the World, 2008). As a result, pesticide residues in vegetables are among the most common causes of food poisoning in China (e.g. Deng *et al.*, 2003; Li, 2002).

To reduce the risk of pesticide residues, social science researchers are conducting studies on farmers adoption of low-toxic or biological pesticides, as alternatives to highly toxic pesticides, as recommended by the Chinese government. Zhang *et al.* (2004) conducted a survey in 15 counties of Shanxi, Shaanxi and Shandong Provinces and empirically analyzed the factors which affected farmers' adoption of non-polluting and green pesticides[1]. They found that perceptions of pesticides depended on contracts with food processing enterprises and joint specialized farmer cooperatives which positively affected the application of non-polluting and green pesticides, while a farmer's house-

---

[1] To improve international competition in Chinese agricultural products and to ensure the health of domestic consumers, the Chinese government conducts authentication work which authenticates food into non-polluted food, green food and organic food. Non-polluting pesticides are those selected and recommended by the National Agricultural Technical Extension and Service Center (NATESC) which is in charge of the authentication of non-polluted food. Green food, however, is overseen by the China Green Food Development Center (CGFDC) and the green pesticides are recommended by the Pesticide Application Guideline for Green Food Production published by the CGFDC. Both the non-polluting pesticides and the green pesticides can be characterized by low toxicity, this study therefore, combines these two types of pesticides as low-toxic pesticides, compared with the highly toxic pesticides not selected and not recommended by either of the centers.

hold size was identified as a negative factor. On the other hand, Huang (2005) qualitatively analyzed the problems of the adoption of biological pesticides in China and found that high prices, low effectiveness and difficulties in application were the main obstacles.

Studies have also been carried out from a scientific perspective in order to decrease or remove pesticide residues from agro-produce. For example, to understand how daily food preparation procedures influence pesticide residues in cabbage, Zhang *et al.* (2007) investigated the effects of washing with tap water, measuring various concentrations of sodium chloride solution or acetic acid solution, refrigeration and cooking for different lengths of time on pesticide residues in cabbage using gas chromatography. Wu *et al.* (2007) tried to remove residual pesticides from vegetables using ozonated water to avoid the adverse impact of these residues on human health.

Unfortunately, despite these attempts to find effective ways of preventing pesticide residues in vegetables, great challenges still remain to ensure the safety of vegetables in China. In fact, during our survey we found that highly toxic pesticides, which are more than likely to cause high pesticide residues in vegetables (Zhang *et al.*, 1999; Zhang *et al.*, 2004), are used in vegetable production by vegetable farmers[2]. Thus, we argue that identification of the risks associated with farmers using highly toxic pesticides and controlling the application of pesticides are extremely important in China, as this will help to prevent vegetable safety issues in the future. To the best of our knowledge, no research has been conducted to address this issue directly. Based on a survey of 507 vegetable farmers in Zhejiang Province, China, this study therefore uses a social science perspective to identify and control vegetable farmers who risk spraying highly toxic pesticides onto vegetables.

2.2 Method

We used a questionnaire-based personal interview technique to collect first

---

[2] Zhang *et al.* (1999) also reported that highly toxic pesticides are not only extensively used in Huaibei in Anhui Province but are also abused by vegetable farmers.

hand data as used in previous publications (e.g. Jin *et al.*, 2008). A draft of the questionnaire was first developed based on existing publications in 2004. 20 vegetable farmers in Hangzhou city, the capital of Zhejiang Province, were selected for the pre-test. We and postgraduate students at the Center for Agriculture and Rural Development, Zhejiang University, interviewed the sample of vegetable farmers. Useful information on question content, wording, sequence, form, layout and question difficulty was collected to improve the draft questionnaire.

The final questionnaire was made up of three sections. Questions in the first section related mainly to the demographic information concerning the vegetable farmers interviewed. The second section consisted of questions on the pesticides currently used, marketing channels for their vegetable products and some other questions. The third section consisted of questions on the farmers' perceptions of vegetable safety issues.

Interviewers for field work were recruited from undergraduate and postgraduate students majoring in agricultural economics and management at Zhejiang University. Social acceptability of vegetable farmers ensured the quality of data obtained. The interviewers were trained and sent to 10 cities[3] throughout Zhejiang Province. During the survey, the interviewers were also supervised by telephone or email. A total of 507 valid questionnaires were collected. Table 2.1 illustrates the number of valid questionnaires received from each city in Zhejiang Province.

Table 2.1 Valid questionnaires received from 10 cities

Research site	Valid questionnaires
Hangzhou	56
Ningbo	44
Wenzhou	35
Jiaxing	31

(To be continued)

[3] Zhejiang Province is actually made up of 11 cities, Hangzhou, Ningbo, Wenzhou, Jiaxing, Huzhou, Shaoxing, Jinhua, Quzhou, Taizhou, Lishui and Zhoushan. We excluded Zhoushan city as it is an island and was difficult for us to access. We do not think this decision affected the overall result.

(Table 2.1)

Huzhou	33
Shaoxing	99
Jinghua	96
Quzhou	30
Taizhou	52
Lishui	31
Total	507

2.3 Results and Discussions

2.3.1 Demographic Analysis

In order to conduct a quantitative analysis, information on the interviewed vegetable farmers was collected using open-ended questions. Table 2.2 summarizes the demographics of the interviewed vegetable farmers[4].

2.3.1.1 *Age*

The average age of the vegetable farmers was 49.66 years, the youngest was 28 years and the oldest was 79 years. About half (43.6%) of the vegetable farmers were 46 to 55 years of age.

2.3.1.2 *Education Level*

The average number of years of education received by the vegetable farmers was 5.35 years, with a standard deviation of 2.87. Generally, the education level of the farmers was low. 46 respondents were illiterate, which accounted for 9.1% of the total and the best educated respondent was a college graduate with 14 years of education.

---

[4] We did not report the gender of the respondents because almost all of the vegetable farmers interviewed were male.

Table 2.2 Demographic characteristics of the interviewed vegetable farmers

Characteristics		Frequency	% of total
Age (years)	< 36	25	4.9
	36 – 45	134	26.4
	46 – 55	221	43.6
	56 – 65	105	20.7
	> 65	22	4.3
Education (years)	0	46	9.1
	1 – 5	216	42.6
	6 – 10	222	43.8
	> 10	23	4.5
Planting area (ha)	< 0.33	217	42.8
	0.33 – 1	248	48.9
	> 1	42	8.3
Household size (number)	< 4	15	3.0
	4 – 6	327	64.5
	> 6	165	32.5
Professional years of farming (years)	< 10	262	51.7
	10 – 20	183	36.1
	> 20	62	12.2
Cooperative membership	Yes	284	56.0
	No	223	44.0
Training received	Frequently	290	57.4
	Seldom	117	23.2
	Never	98	19.4

2.3.1.3 *Planting Area*

The planting areas varied from 0.1 ha to 20 ha, with a mean of 9.27 and a standard deviation of 18.21. The three biggest planting areas were 20 ha, 13.33 ha and 9.87 ha, and 42 vegetable farmers had farms of more than 1 ha. However, most of the vegetable farmers (42.8%) could be characterized as small scale farmers with planting areas of less than 0.33 ha.

2.3.1.4 *Household Size*

The average household size was 4.05 members (standard deviation 1.20). The most common household size was 4–6 members, which was 64.5% of the total sample.

2.3.1.5 *Professional Years of Farming*

On average, the respondents had been engaged in vegetable production for 13.07 years (standard deviation 8.34), with a minimum of one year and a maximum of 50 years. Most vegetable farmers had been farming for a number of years and had cumulated experience in vegetable production.

2.3.1.6 *Cooperative Membership*

284 (56%) of the vegetable farmers had joined specialized farmers cooperatives. In general, vegetable farmers in Zhejiang Province were well organized.

2.3.1.7 *Training Received*

Most (290 or 57.4% of total) vegetable farmers claimed that they had received frequent training in vegetable production. However, there were 98 vegetable farmers who had never attended training sessions.

The main marketing channels for the farmers interviewed are shown in Fig. 2.1. Wholesale markets were selected as the most important places to sell vegetables followed by handlers and processing enterprises. The vegetable farmers were also asked to indicate the amount of vegetables produced for self consumption, which was calculated as a mean of 3.4%, with a standard deviation of 8.42.

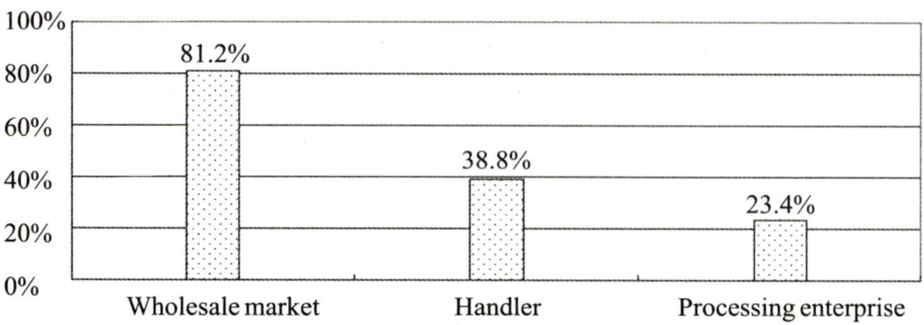

Fig. 2.1   Marketing channels of interviewed vegetable farmers

2.3.2 Pesticides Used

Understanding the types of pesticides used by vegetable farmers is not only critical to ensuring the success of this study but is also of great interest. In order to obtain exact answers with regard to the pesticides used, we did not ask the respondents directly whether they were using highly toxic pesticides, but listed all the pesticides that may be adopted by vegetable farmers and asked them to select the pesticides they were currently using. The reasons for this were two-fold. Firstly, if we had asked directly, we may not have received accurate answers, as the world of highly toxic pesticides is thought to be very sensitive to vegetable farmers. Secondly, farmers may not have been able to tell whether the pesticides they were using were highly toxic or not. The results of this investigation are depicted in Fig. 2.2, which show that as many as 121 (23.9%) of the 507 vegetable farmers used highly toxic pesticides.

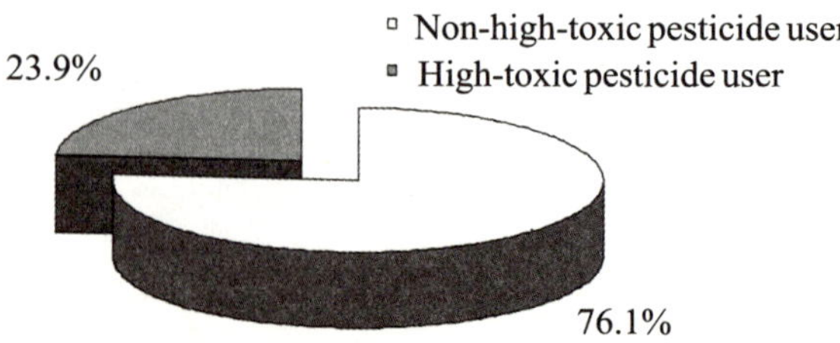

Fig. 2.2 Pesticides used by interviewed vegetable farmers

2.3.3 Perceptions of Vegetable Safety Issues

Results of the 8 questions on perceptions of vegetable safety issues are listed in Table 2.3. The results of questions 7 and 8 deserve a special mention. Respondents attributed much more importance to the appearance of their vegetables rather than their safety status when selling their produce. Thus the appearance of vegetables may be an important reason why vegetable farmers use pesticides extensively throughout China.

Table 2.3 Perceptions of vegetable safety issues

1. I care about production information to ensure vegetable safety:					
Yes, exactly				No	
1	2	3	4	5	Total
53	290	122	37	4	506
10.5%	57.3%	24.1%	7.3%	0.8%	100.0%

2. The vegetable safety situation is serious nowadays:					
Yes, exactly				No	
1	2	3	4	5	Total
39	136	182	131	17	505
7.7%	26.9%	36.0%	25.9%	3.4%	100.0%

3. Recognition of three types of vegetables[5]:					
Bad				Good	
1	2	3	4	5	Total
92	93	138	126	58	507
18.1%	18.3%	27.2%	24.9%	11.4%	100

(To be continued)

---

[5] Three kinds of vegetables here are non-polluted vegetables, green vegetables and organic vegetables, as mentioned in footnote 1. Recognition is tested by asking 6 different questions. In the first three questions, we provided the markets with non-polluted food, green food and organic food, and asked the respondents whether they had ever seen the foods or not (0 for had not, 1 for had). We listed non-polluted vegetables, green vegetables and organic vegetables in the next three questions and asked the respondents whether they had ever heard of these or not (0 for had not, 1 for had). The scores were then summed up into a variable ranging from 0 to 7. Here we merged score 2 with 3 and score 5 with 6, and arranged them into a 5 point Likert-type scale in order to match the scales of other questions.

(Table 2.3)

4. It is of great significance to obtain vegetable certifications:					
Yes, exactly				No	
1	2	3	4	5	Total
69	237	101	86	14	507
13.6%	46.7%	19.9%	17.0%	2.8%	100.0%
5. I can increase revenue by producing safer vegetables:					
Yes, exactly				No	
1	2	3	4	5	Total
62	269	99	70	6	506
12.3%	53.2%	19.6%	13.8%	1.2%	100.0%
6. I follow other vegetable farmers around me to ensure vegetable safety:					
Yes, exactly				No	
1	2	3	4	5	Total
28	191	90	178	19	506
5.5%	37.7%	17.8%	35.2%	3.8%	100.0%
7. The better the vegetable's appearance, the easier it is to sell:					
Yes, exactly				No	
1	2	3	4	5	Total
239	245	19	3	0	506
47.2%	48.4%	3.8%	0.6%	0.0%	100.0%
8. The safer the vegetables is, the easier it is to sell:					
Yes, exactly				No	
1	2	3	4	5	Total
89	247	93	72	1	502
17.7%	49.2%	18.5%	14.3%	0.2%	100.0%

2.3.4 Identification of Farmers at Risk of Using Highly Toxic Pesticides

In this section, we divided the vegetable farmers into two groups according to their pesticide usage (whether or not they used highly toxic pesticides in vegetable production). The t-statistic was employed to compare the characteristics of the two groups of farmers and to identify the farmers at risk of using highly toxic pesticides. Table 2.4 shows the statistical results.

In general, highly toxic pesticide users could be characterized as older and less educated vegetable farmers. On average, highly toxic pesticide users were 2.08 years older and 0.52 years less educated than those who did not use highly toxic pesticides. The difference between the two groups in terms of the number of farming years indicated that older farmers were used to applying highly toxic pesticides which had been introduced in previous years and they found it difficult to change.

Table 2.4　At risk farmers' identification results

	Highly toxic pesticides	Mean	Std. Dev.	t-test
Demographic characteristics:				
Age (years)	No	49.17	8.97	-2.243^{**}
	Yes	51.26	8.84	
Education (years)	No	5.47	2.80	1.742^{*}
	Yes	4.95	3.04	
Planting area (ha)	No	0.65	17.96	1.064
	Yes	0.52	18.99	
Household size (number)	No	3.99	1.14	-1.584
	Yes	4.20	1.28	
Professional years (years)	No	12.14	7.60	-3.996^{***}
	Yes	16.03	9.84	
Cooperative membership (1 for yes, otherwise 0)	No	0.62	0.49	4.898^{***}
	Yes	0.37	0.49	
Training received (5 point scale from 0 to 5, with 5 for frequently, 0 for never)	No	2.62	1.26	-4.798^{***}
	Yes	3.31	1.42	
Rate of self consumption (%)	No	2.69	8.00	-3.239^{***}
	Yes	5.74	9.31	
Marketing channels: (1 for yes, otherwise 0)				
Wholesale market	No	0.83	0.37	2.038^{**}
	Yes	0.74	0.44	
Handler	No	0.31	0.46	-6.079^{***}
	Yes	0.62	0.49	
Processing enterprise	No	0.27	0.44	3.388^{***}
	Yes	0.13	0.34	

(To be continued)

(Table 2.4)

Perceptions of vegetable safety issues:
(Question 3 is 7 point scale, others are 5 point scale)

1. I care about production information to ensure vegetable safety.	No	2.25	0.72	−2.206**
	Yes	2.46	0.98	
2. Vegetable safety situation is serious nowadays.	No	2.90	0.97	0.845
	Yes	2.88	1.07	
3. Recognition of three kinds of vegetables.	No	2.98	1.97	2.132**
	Yes	2.55	1.72	
4. It is of great significance to obtain vegetable certifications.	No	2.49	0.97	0.872
	Yes	2.47	1.15	
5. I can increase revenue by producing safer vegetables.	No	2.38	0.87	0.887
	Yes	2.40	1.04	
6. I follow other vegetable farmers around me to ensure vegetable safety.	No	2.99	1.03	2.154**
	Yes	2.76	1.09	
7. The better the vegetable's appearance, the easier it is to sell.	No	1.61	0.59	1.897*
	Yes	1.49	0.61	
8. The safer the vegetable is, the easier it is to sell.	No	2.30	0.92	0.744
	Yes	2.23	1.01	

Note: *, **, *** significant at 10%, 5%, and 1%, respectively.

Unspecialized vegetable farmers were more likely to use highly toxic pesticides than specialized farmers. The results of the statistical t-tests indicated that a higher self-consumption rate resulted in a higher tendency to use highly toxic pesticides. These results were also partly supported by the comparison of the planting areas, where the mean area (0.65 ha) planted by highly toxic pesticide users was larger than that (0.52 ha) planted by those not using highly toxic pesticides, although the t-statistic value for this finding was not significant. The results implied that unspecialized vegetable farmers lack basic knowledge of pesticides, and sprayed highly toxic pesticides on vegetables largely because they did not know the highly toxic nature of these pesticides.

The results also showed that vegetable farmers who received less training had a tendency to apply highly toxic pesticides and cooperative members were less likely to be highly toxic pesticide users. 62% of farmers who did not use highly toxic pesticides were cooperative members, which was in sharp contrast to the 37% of highly toxic pesticide users who joined cooperatives. This implied that vegetable production did not benefit from the technical support provided by specialized farmer cooperatives. This finding is consistent with Wei and Lu (2004), who stressed the importance of specialized farmer cooperatives in controlling and improving the quality of food products based on interviews with farmers and specialized farmer cooperatives in Zhejiang Province.

In addition, understanding the differences in the marketing channels of the two groups is of great importance in controlling problematic vegetables before circulation. It is very interesting to note that the vegetable farmers who used highly toxic pesticides had a high probability of selling their vegetables to handlers, as 62% of the farmers dealt with handlers. Compared with those farmers who were using highly toxic pesticides, farmers who were not using highly toxic pesticides were more likely to sell their vegetables at wholesale markets or to agricultural product processing enterprises. There are two possible explanations for these findings. One is that pesticide residue detecting systems have already been set up in wholesale markets and agricultural product processing enterprises, and high risk occurs when highly toxic pesticides are used during production. However, the handlers do not use pesticide detecting instruments. Another possible explanation is that compared with wholesale markets and agricultural product processing enterprises, the traceability of vegetables collected by the handlers is more difficult as they collect vegetables from a large number of vegetable farmers.

By analyzing perceptions of vegetable safety issues, we found that vegetable farmers who were using highly toxic pesticides could be described as those who cared less about production information to ensure vegetable safety, did not recognize the three types of vegetables, were more likely to follow other vegetable farmers around them to ensure vegetable safety, and attributed more importance to the appearance of the vegetables. Therefore, a poor understanding of vegetable safety issues may be another important reason in explaining why vegetable farmers spray highly toxic pesticides onto vegetables in China.

2.4 Conclusions and Implications

To ensure productivity, pesticides are extensively used in vegetable production in China. However, these pesticides cause residue problems which can damage the health of Chinese consumers and pose problems for international trade. Great challenges still remain to ensure vegetable safety in China, as highly toxic pesticides are widely used in vegetable production, which are more than likely to cause high pesticide residues in vegetables (Zhang *et al.*, 1999; Zhang *et al.*, 2004). In this study, we conducted a survey of 507 vegetable farmers in Zhejiang Province to identify and control vegetable farmers who are at high risk of spraying highly toxic pesticides onto vegetables in China.

The main findings of this study are as follows: Firstly, highly toxic pesticide users can be characterized by older and less educated vegetable farmers. Secondly, unspecialized vegetable farmers were more likely to use highly toxic pesticides than specialized farmers. Thirdly, vegetable farmers who received less training and selected handlers as their marketing channel had a tendency to apply highly toxic pesticides and cooperative members were less likely to be highly toxic pesticide users. Finally, vegetable farmers who had a poor understanding of vegetable safety issues were more likely to use highly toxic pesticides.

These findings suggest that training programs and extension services, which cover current government policies on vegetable safety, provide exact knowledge of safe vegetables and knowledge of highly toxic pesticides should be available, especially for older and less educated vegetable farmers. Another implication of this study is that the government should encourage vegetable farmers to join cooperatives. These cooperatives may contribute to the quality improvement of vegetables by providing on-site training in use of pesticides for their members. More importantly, it is necessary to promote specialized vegetable farmers. Priority policies should be given to vegetable farmers to expand their planting areas, as specialized farmers have more knowledge of vegetable safety and can control the use of highly toxic pesticides. In addition, pesticide residue detecting systems should be set up to ensure the quality of vegetables collected by handlers.

References

Calvin, L., Gale, F., Hu, D. & Lohmar, B. (2006). Food safety improvements underway in China. *Amber Waves*, 4(5),16-21.

Chen, C., Yang, J. & Findlay, C. (2008). Measuring the effect of food safety standards on China's agricultural exports. *Review of World Economics*, 144, 83-106.

China Statistical Yearbook (2007). National Statistical Bureau of China. Beijing: China Statistics Press.

Deng, L., Qu, H., Huang, R., Yang, Y., Zheng, X. & Wang, H. (2003). Survey of food poisoning by organosphorus pesticide at an employee refectory. *Practical Preventive Medicine*, 10(5), 766-767.

Huang, C. (2005). Negative externalities of pesticide use and the economic analysis. *Journal of Anhui Agricultural Sciences*, 1, 151-153.

Huang, J., Hu, R. Pray, C., Qiao, F. & Rozelle, S. (2003). Biotechnology as an alternative to chemical pesticides: a case study of Bt cotton in China. *Agricultural Economics*, 29, 55-67.

Jin, S., Zhou, J. & Ye, J. (2008). Adoption of HACCP system in the Chinese food industry: A comparative analysis. *Food Control*, 19, 823-828.

Li, X. (2002). Analysis of food poisoning due to taking vegetables contaminated with organophosphorus pesticide. *China Tropical Medicine*, 2(4), 519.

Ngowi, A.V.F., Mbise, T.J., Ijani, A.S.M., London, L. & Ajayi, O.C. (2007). Smallholder vegetable farmers in Northern Tanzania: Pesticides use practices, perceptions, cost and health effects. *Crop Protection*, 26, 1617-1624.

Statistics of the World (2008). Statistics Bureau and the Director-General for Policy Planning (Statistical Standards) & Statistical Research and Training Institute of Ministry of Internal Affairs and Communications of Japan. Beijing: Statistics Bureau Press.

Wei, L. & Lu, G. (2004). Functions of farmer specialized cooperatives on agro-products quality control: Case studies of several farmer specialized cooperatives in Zhejiang University. *Chinese Rural Economy*, 2, 36-41.

Widawsky, D., Rozelle, S., Jin, S. & Huang, J. (1998). Pesticide productivity, host-plant resistance and productivity in China. *Agricultural Economics*, 19, 203-217.

Wu, J., Luan, T., Lan, C. Lo, H. & Chan, G. (2007). Removal of residual pesticides on vegetables using ozonated water. *Food Control*, 18, 466-472.

Zhang, C., Chen, C. & Li, H. (1999). Pesticide application in vegetable production: Current situation and countermeasure. *Vegetables*, 3, 18-19.

Zhang, Z., Liu, X. & Hong, X. (2007). Effects of home preparation on pesticide residues in cabbage. *Food Control*, 18, 1484-1487.

Zhang, Y., Ma, J., Kong, X. & Zhu, Y. (2004). Factors that affect farmers' adoption of non-pollution and green pesticides: Empirical analysis based on data from 15 counties (cities) in Shanxi, Shaanxi and Shandong Province. *Chinese Rural Economy*, 1, 41-49.

3

Adoption of Food Safety and Quality Standards by China's Agricultural Cooperatives

3.1 Introduction

In recent years, China has frequently experienced food safety scares due to problems related to pesticide residues. Recently, the government established a nationwide food inspection and monitoring system (Wang *et al.*, 2008) that involves recording and publicizing the quality of vegetables in 37 major cities[1] based on food standards derived from the joint FAO/WHO Codex Alimentarius Commission (CAC) international standards. As a result, the proportion of vegetables of acceptable quality has increased from 82% in 2003 to

---

[1] Beijing, Changchun, Changsha, Chengdu, Chongqing, Dalian, Fuzhou, Guangzhou, Haerbin, Haikou, Hangzhou, Hefei, Huhehaote, Jinan, Kunming, Lanzhou, Lasa, Nanchang, Nanjing, Nanning, Ningbo, Qingdao, Guiyang, Shanghai, Shenyang, Shenzhen, Shijiazhuang, Shouguang, Taiyuan, Tianjin, Wuhan, Urumqi, Xiamen, Xining, Xi'an, Yinchuan, Zhengzhou.

94% in 2007 in terms of being free of pesticide residues. Food inspection and the monitoring system have contributed significantly to improving food safety in China.

Nevertheless, a food inspection and monitoring system that is conducted by testing the end-products is limited in its ability to assess food safety. Although the rate at which the quality of vegetables has increased between 2004 and 2007, it has never surpassed a level of 95%. We argue that in addition to testing, it is also important to standardize the production practices used by farmers since there have been frequent reports of the abuse of pesticides during production (e.g. Zhou, 2005). Currently, the available Chinese domestic standards in agricultural production mainly include pollution-free food standards, green food standards and organic food standards which have been proposed by the National Agricultural Technical Extension and Service Center (NATESC), the China Green Food Development Center (CGFDC) and the China Organic Food Certification Center (COFCC), respectively[2]. The Ministry of Agriculture is the lead agency promoting food safety at the farm level by encouraging the adoption of domestic food safety and quality standards[3]. But, as pointed out by Calvin *et al.* (2006), it is difficult to standardize production practices in a sector composed of 200 million farm households who typically have 1–2 acres of land divided into 4–6 noncontiguous plots, as is the case in China today. The reasons are twofold. On the one hand, most small-scale farmers cannot afford the costs associated with the implementation of these standards (Han, 2007). On the other hand, the majority of farmers are not well educated and do not fully understand the key points of standardization[4].

Agricultural cooperatives and other arrangements for aggregating produc-

---

[2] Although international food safety and quality controls (e.g. Good Agricultural Practices) are also available in China, they are rarely adopted by agri-food producers due to their high implementation costs.

[3] From 2008, the Ministry of Agriculture has been discussing plans to implement mandatory programs to encourage the adoption of the Pollution Free Standards.

[4] Zhou (2005) found that vegetable farmers in Zhejiang Province received an average education of only 5.43 years.

ers into larger "production bases" facilitate the main adopters of food quality and safety standards in China. In fact, encouraging the adoption of these standards by agricultural cooperatives is a more practical and feasible alternative to regulations aimed at farmers[5], because not only can the related costs be shared by a group of small-scale farmers but the production practices can also be organized by the cooperatives. Understanding the mechanism of the adoption of food quality and safety standards by agricultural cooperatives[6] is of great importance in standardizing agri-food production practices in China.

The overall objective of this study is to analyze the factors that affect the adoption of food safety and quality standards in China based on data from a survey of 124 vegetable cooperatives. The vegetable sector has been chosen because ensuring the quality of vegetables is of extreme importance in China as vegetables are the major constituent of the Chinese diet. More importantly, toxic pesticide residues on vegetables are one of China's most pressing food safety concerns and many cooperatives have been formed to produce and market vegetables.

This chapter is organized as follows. Following this introduction, an empirical model is introduced along with data description. The results and discussion are then provided in Section 3.4 and are followed by policy implications and conclusions.

3.2 Methodology

A cooperative has two alternatives, either to adopt or ignore any food safety and quality standards. We assume a cooperative's utility resulting from either alternative depends upon several attributes of the cooperative. The utility of an alternative is a function of the attributes of the cooperative, which is given

---

[5] Existing studies (e.g. Wei and Lu, 2004; Ren and Ge, 2008) stress the importance of specialized farm cooperatives in controlling and improving the quality of food products based on interviews to farmers and farmer-specialized cooperatives in Zhejiang Province.

[6] The existing literature on the adoption of food quality and safety standards has mostly been conducted in developed countries, the findings of which may not be applicable in developing countries like China.

by

$$U_0^* = \beta'_0 x + \varepsilon_0 \tag{3.1}$$

where U_0^* is the utility of choosing an alternative; x is a vector containing the attributes of the cooperative; β_0 is a parameter vector, and ε_0 is the error term which allows for the uncertainty.

The utility of adopting food safety and quality standards can then be specified as

$$U_A^* = \beta'_A x + \varepsilon_A \tag{3.2}$$

where U_A^*, β_A, and ε_A are the utility, parameter vector and the stochastic function of adopting food safety and quality standards, respectively.

If the cooperative does not adopt food safety and quality standards, we have

$$U_N^* = \beta'_N x + \varepsilon_N \tag{3.3}$$

where U_N^*, β_N, and ε_N are the utility, parameter vector and the stochastic function of not adopting food safety and quality standards, respectively.

Therefore, the cooperative's net utility between adopting and ignoring is

$$\begin{aligned} U^* &= U_A^* - U_N^* \\ &= (\beta'_A - \beta'_N)x + (\varepsilon_A - \varepsilon_N) \\ &= \beta'x + \varepsilon \end{aligned} \tag{3.4}$$

where U^*, β, and ε are the net utility, parameter vector to be estimated and the stochastic function, respectively.

As the cooperative's net utility is a latent variable, we cannot observe it directly. But if $U^* > 0$, the observed choice will be the adoption of food safety and quality standards (or *Adoption* = 1) and if $U^* \leq 0$, the observed choice will be the non-adoption of food safety and quality standards *Adoption* = 0.

$$Adoption = \begin{cases} 1, & U^* > 0 \\ 0, & U^* \leq 0 \end{cases} \tag{3.5}$$

If we assume the stochastic function ε follows a logistic distribution with mean 0, and a variance of $\pi^2/3$, the probabilities of *Adoption* = 1 or 0 can be expressed as

$$P(Adoption = 1) = P(U^* > 0)$$
$$= P(\varepsilon < \beta'x)$$
$$= \frac{1}{1 + e^{-\beta'x}} = \Lambda(\beta'x) \tag{3.6}$$

$$P(Adoption = 0) = P(U^* \leq 0)$$
$$= P(\varepsilon \geq \beta'x)$$
$$= 1 - \frac{1}{1 + e^{-\beta'x}} = 1 - \Lambda(\beta'x) \tag{3.7}$$

The likelihood function can be written as

$$L = \prod [\Lambda(\beta'x)]^{Adoption} [1 - \Lambda(\beta'x)]^{1 - Adoption} \tag{3.8}$$

The parameter vector β in Eq. (3.8) can be estimated using the maximum likelihood method. The marginal effect for a variable x_i can be calculated as follows:

$$\frac{dP}{dx_i} = \Lambda(\beta'x)[1 - \Lambda(\beta'x)]\beta_i \tag{3.9}$$

3.3 Data Source and Variable Description

The data for our empirical study was collected from 10 districts throughout Zhejiang Province[7]. Based on a list provided by the Department of Agriculture of Zhejiang Province, we sent questionnaires to 270 vegetable cooperatives randomly selected in Zhejiang Province. In total, 124 valid questionnaires were returned during the period September 2006 to March 2007. Fig. 3.1 illustrates the number of valid questionnaires received from each city across Zhejiang Province.

---

[7] In fact, Zhejiang Province is made up of 11 districts, Hangzhou, Ningbo, Wenzhou, Jiaxing, Huzhou, Shaoxing, Jinhua, Quzhou, Taizhou, Lishui and Zhoushan. We excluded Zhoushan city as it is an island and we do not think the overall result is less valid based on this decision.

Fig. 3.1 Valid respondents from each district in Zhejiang Province

Table 3.1 lists the variables used in empirical analysis. The dependent variable is the dichotomous outcome of whether or not to adopt food safety and quality standards. According to the results of the survey, 78.2% of the vegetable cooperatives have adopted food safety and quality standards. Since Zhejiang is a rapidly developing province, the rate of adoption of the food safety and quality standards in this province is higher than the average national rate.

We hypothesise that the following attributes affect the adoption decision: (1) cooperative size, (2) innovativeness, (3) perception, (4) reputation, (5) cost and benefit, (6) price premium, (7) customer attraction, (8) destination market, (9) support[8] (see Table 3.1). The independent variables are specified

---

[8] Although meeting mandatory requirements is reported to be a key factor in the adoption decision in empirical studies conducted in developed countries (e.g. Henson and Holt, 2000; Fouayzi et al., 2006;), we have not included this in our analysis because implementing food safety and quality standards is voluntary in China.

Table 3.1 Descriptive statistics of variables

		Mean	Min.	Max.	Std. Dev.
Adoption	=1 if any standard is adopted, =0 if no standard is adopted	0.782	0	1	0.414
Cooperative size	Land size of cooperative (ha)	154.160	1.33	1266.67	231.921
Innovativeness	=1 if cooperative has a homepage, =0 otherwise	0.427	0	1	0.497
Perception	5 point Likert type scale from fully disagree (= 1) to fully agree (= 5) with the statement that implementing a standard will improve the quality of the vegetables	4.371	1	5	0.738
Reputation	=1 if cooperative has a brand name, =0 otherwise	0.726	0	1	0.448
Cost and benefit	=1 if the expected benefit covers the costs of implementing a standard, =0 otherwise	0.589	0	1	0.494
Price premium	=1 if the cooperative expects a price premium by implementing a standard, =0 otherwise	0.645	0	1	0.480
Customer attraction	=1 if the standard is helpful in attracting customers, =0 otherwise	0.935	0	1	0.247
Destination market	=1 if the cooperative supplies supermarkets or foreign markets, =0 otherwise	0.266	0	1	0.444
Availability of support	=1 if cooperative is supported by downstream buyers, =0 otherwise	0.323	0	1	0.469

for two reasons. Firstly, factors mentioned in previous studies are considered. By including these factors, we can compare the influence of our factors on the adoption decision regarding food quality and safety with that of the factors described in previous studies. Secondly, because much of the existing literature on the adoption of food quality and safety standards has been conducted in developed countries, the findings of which may not be fully applicable to developing countries such as China, we added three other independent variables. Specifically, we include (3) perception, (6) price premium and (9) support in order to better understand the adoption decision in China. The variables are explained in greater detail as follows.

(1) Cooperative size. The size of the firm has been identified as one of the most important factors to affect the adoption decision. Jayasinghe-Mudalige and Henson (2007) argued that larger firms have the capacity to implement food safety controls while most small firms showed no desire to do so due to their smaller resources. The size of the firm is therefore expected to have a positive effect on the implementation of food safety standards.

(2) Innovativeness. Innovativeness has also been discussed in previous studies (e.g. Herath *et al.*, 2007). We have identified the presence of a homepage for a cooperative as an indicator of innovativeness and expect this to be positively related to the uptake of food safety standards.

(3) Perception. A major barrier to the adoption of food safety procedures in developing countries may be the level of available knowledge relating to food quality and safety standards. We used a 5-point Likert scale that ranged from 'fully disagree' to 'fully agree' with the statement that implementing a standard will improve the vegetable quality, in order to measure the respondent's perception about food quality and safety standards. A high score should lead to a high probability of the adoption of safety standards.

(4) Reputation. Reputation will generate benefits as a result of consumers' repeat purchases and customer loyalty. On the other hand, it will also bring about devastating losses in the event of a food-related accident (Herath *et al.*, 2007). Whether or not the cooperative has a brand name is used as a proxy for the effect of reputation in this chapter. The use of a brand name is expected to

be positively related to the adoption of food safety procedures.

(5) Cost and benefit. In previous studies, the firm's expected benefit has also been discussed as an important factor in the decision to adopt safety and quality standards. For instance, Holleran *et al.* (1999) indicated that if the benefits of certification of a quality assurance system exceed the adoption and maintenance costs, then the standard is worthwhile. Due to the lack of an educated workforce and the necessary equipment required to implement many of the standards, adopting food quality and safety standards is a considerable financial burden for vegetable cooperatives in China. Vegetable cooperatives will adopt food quality and safety standards if the expected benefit covers the associated costs.

(6) Price premium. The purpose of integrating this variable into the empirical analysis is to test the hypothesis that cooperatives will make a positive adoption decision if they expect to achieve a price premium by implementing a particular standard. Thus, a price premium is expected to be positively related to the adoption of standards.

(7) Customer attraction. If the respondents view the implementation of food quality and safety standards as a strategy to attract new customers, the possibility of adoption will increase.

(8) Destination market. Consumer demand for higher quality produce and the uptake of standards by firms would be expected to be greater if the cooperative serves domestic supermarkets or foreign markets. As stated by Jayasinghe-Mudalige and Henson (2007), many supermarket chains and food service operators in North America require their suppliers to adopt specific food safety controls. The probability of adoption increases if the destination market is a supermarket or foreign market.

(9) Availability of support. As mentioned above, adopting standards is a financial and practical burden for many cooperatives and in some cases downstream members may provide support to cooperatives. The adoption decision is more likely to be positive when support is available.

The above nine attributes are used in the logistic model to investigate the adoption decision regarding food quality and safety standards by vegetable cooperatives in China. The attributes (1) through (3) are related to the characteristics of the cooperative, and attributes (2) through (7) are associated with internal factors; (8) and (9) are external factors.

3.4　Results and Discussions

Table 3.2 shows the statistical results of the logistic model analysis. Generally, the model performs well, with a McFadden Pseudo R^2 value of 0.359 and log likelihood value of -41.620. In total, 86.3% of adoption decisions were correctly predicted.

Table 3.2　Statistical results for adoption decision

	Coefficient	Std. Error	Pr. $> |z|$	Marginal Pr.
Intercept	-6.713	2.043	0.001	-
Cooperative size	0.010	0.004	0.032	0.001
Innovativeness	0.076	0.592	0.898	0.011
Perception	0.934	0.448	0.037	0.137
Reputation	1.839	0.631	0.004	0.270
Cost and benefit	1.060	0.575	0.065	0.155
Price premium	0.803	0.645	0.213	0.118
Customer attraction	0.957	1.343	0.476	0.140
Destination market	1.639	0.891	0.066	0.240
Availability of support	-0.849	0.621	0.171	-0.125
McFadden Pseudo R^2	0.359			
Log likelihood	-41.620			
Correct predictions	86.3%			
Observations	124			

The relationship between cooperative characteristics and the adoption decision is explored first. Cooperative size, approximated by the land area of the cooperative, is an important factor that affects the adoption decision as its coefficient is positively signed and statistically significant. This result indicates that economies of size exist in the adoption of food quality and safety standards by Chinese agricultural cooperatives. A positive effect is found for the innovativeness variable measured in terms of whether or not the cooperative possesses a homepage, although its lack of statistical significance is disappointing. Our result only partially supports the finding by Herath et al. (2007), that innovativeness is positively associated with the adoption of food safety and quality controls in the Canadian food processing sector. A possible reason for this may be that the method of measurement used in our study is different from that in Herath *et al.* (2007)[9].

A positive perception of food quality and safety standards is found to be one of the most important factors affecting the adoption decision. It seems that doubt about the effectiveness of food quality and safety standards in ensuring vegetable quality is a major obstacle to the adoption decision in China. This result is quite common in developing countries. For example, in a study on the adoption of the Euro gap standard by mango producers in Peru, Kleinwechter and Grethe (2006) reported that access to information about the standard is a major barrier to its adoption.

Turning to the results of the attributes related to internal factors, reputation, which was measured by whether or not a firm owned a brand, is the most important factor affecting the adoption decision. Based on the estimated marginal effect, the probability of a cooperative adopting a food quality and safety standard increases by 27% if it has a brand. This may indicate that once a cooperative has a registered brand, it will pay more attention to the quality of its vegetables and vice versa. As expected, there was a positive and statistically significant relationship between the expected profit and the adoption decision, which is in accord with previous studies (e.g. Holleran *et al.*, 1999; Hen-

---

[9] Herath *et al.* (2007) measured innovativeness by whether food processing firms had adopted at least one innovation during the period from 1995 to 1997.

son and Holt, 2000; Fouayzi *et al.*, 2006). The cooperatives are rational and will not adopt the standard if it is not worthwhile. However, to our surprise, the adoption decision was not influenced by either price premium or customer attraction. A possible reason for this result is that the market for vegetables produced under standards is currently in chaos as there are many counterfeits in China. The cooperatives may not be able to obtain a price premium or attract customers by labeling food quality and safety standards.

Destination market and support from downstream members were tested as external factors. A positive and statistically significant effect is found for the destination market variable, which is approximated by whether the cooperative serves supermarkets or foreign markets. The marginal effect indicates that the possibility of adopting a food quality and safety standard increases by 24% if a cooperative deals with supermarkets or exports its vegetables to foreign countries. In general, our result is in accord with previous studies by Holleran *et al.* (1999), Henson and Holt (2000), Fouayzi *et al.* (2006) , Jayasinghe-Mudalige and Henson (2007), Wang *et al.* (2009), but is not consistent with the findings of Herath *et al.* (2007) who reported that the adoption of enhanced food safety practices in the Canadian food processing sector cannot be fully explained by the maintenance and/or improvement of access to foreign markets. This result may have two explanations. A positive relationship between the other external factor, support from downstream members, and the adoption of food quality and safety standards was not confirmed in our study, which may indicate that the cooperatives do not implement a standard merely because of the availability of support.

3.5 Policy Implications and Conclusions

There have been frequent food safety scares in China in recent years. Adoption of food safety and quality standards by China's agricultural cooperatives serves as an important approach for monitoring production practices of the numerous small-scale farmers and thus ensuring food quality in the products produced by them. Based on survey data from 124 vegetable cooperatives in Zhejiang Province, the overall goal of this study was to analyze the factors that affect the adoption of Chinese domestic standards in agricultural production, namely a non-pollution standard, green standard and organic standard by vegetable cooperatives in China.

Based on previous studies, nine factors such as cooperative size, innovativeness, perception, reputation, cost and benefit, price premium, customer attraction, destination market and support are expected to affect the adoption behavior of vegetable cooperatives in China. We analyzed the effect of these factors by using a logistic model. We found that cooperative size, perception of standards, reputation, expected cost and benefit and destination market have a positive and statistically significant relationship with the adoption decision. The effects of the other factors on adoption decisions were not confirmed in our study.

Our results emphasize the importance of fostering the development of agricultural cooperatives in China, especially in terms of land size and brand registration in facilitating food safety and quality standard adoption. Clearly, with the enlargement of the cooperative size, adopting food quality and safety standards will become more affordable for agricultural cooperatives. Also, agricultural cooperatives are likely to treat the quality of the agri-food they provide more seriously once they register a brand name for their products. As such, the possibility of adopting a food quality and safety standard to ensure food safety will increase.

Another implication from our qualitative analysis is that it is also important to provide adequate information to agricultural cooperatives about the ability of food quality and safety standards to ensure the quality of agri-food products. In a developing country like China, agricultural cooperatives may not yet be fully aware of the effectiveness of the food quality and safety standards, and this poses a barrier to their adoption decision.

Finally, our results show that although in general the cooperatives are rational in their decisions regarding whether or not to adopt a food quality and safety standard, it seems that their adoption decision for a food quality and safety standard is not motivated by a desire to attract more customers or to achieve a price premium. We conclude that meeting the requirement of destination markets in order to retain access and, perhaps, maintain the market share is the main incentive for cooperatives to adopt food quality and safety standards in China today. Our results indicated that the destination market (supermarkets or foreign markets) is one of the most important factors affecting the cooperative's adoption decision. This result is in line with the available literature. As also pointed out by Holleran *et al.* (1999), a single benefit, such as satisfying a customer requirement, may be of such importance that the other costs of implementing a quality assurance system become irrelevant. This may be especially true in a developing country like China. As such, encouraging the development of supermarkets and chain store operations in the agri-food retail sector will undoubtedly improve the adoption rate of food quality and safety standards in China[10]. Nevertheless, as argued in Mainville *et al.* (2005), the mechanism of the retailers' decision to use public or private grades and standards needs to be explored in the future. Last but not least, other than cooperatives, agricultural technology extension stations and agricultural enterprises are also encouraged by the Chinese government to standardize the production practices used by small-scale farmers. It is imperative that research be undertaken to understand the mechanism of their adoption of food safety and quality standards.

References

Calvin, L., Gale, F., Hu, D. & Lohmar, B. (2006). Food safety improvements underway in China. *Amber Waves*, 4(5), 16-21.

Fouayzi, H., Caswell, J.A. & Hooker, N.H. (2006). Motivations of fresh-cut produce firms to implement quality management systems. *Review of Agricultural Economics*, 28(1), 132-146.

---

[10] From 2008, the Ministry of Commerce together with the Ministry of Agriculture started a pilot program to encourage supermarket chains to purchase directly from cooperatives or production bases.

Han, S. (2007). Development of rural cooperative economy and organization in Weifang City, Shandong Province. *Chinese Rural Economy*, 8, 56-63.

Henson, S. & Holt, G. (2000). Exploring incentives for the adoption of food safety controls: HACCP implementation in the UK dairy sector. *Review of Agricultural Economics*, 22(2), 407-420.

Herath, D., Hassan, Z. & Henson, S. (2007). Adoption of food safety and quality controls: Do firm characteristics matter? Evidence from the Canadian food processing sector. *Canadian Journal of Agricultural Economics*, 55(3), 299-314.

Holleran, E., Bredahl, M.E. & Zaibet, L. (1999). Private incentives for adopting food safety & quality assurance. *Food Policy*, 24(6), 669-683.

Jayasinghe-Mudalige, U. & Henson S. (2007). Identifying economic incentives for Canadian red meat and poultry processing enterprises to adopt enhanced food safety controls. *Food Controls*, 18(11), 1363-1371.

Kleinwechter, U. & Grethe, H. (2006). The adoption of the Eurepgap standard by mango exporters in Piura, Peru. Contributed Paper presented at the 26th International Association of Agricultural Economists Conference, 12–19 August, Gold Coast, Queensland, Australia.

Mainville, D.Y., Zylbersztajn, D., Farina, E.M. & Reardon, T. (2005). Determinants of retailers' decisions to use public or private grades and standards: evidence from the fresh produce market of Sao Paulo, Brazil. *Food Policy*, 30(3), 334-353.

Ren, G. & Ge, Y. (2008). An analysis on the mechanism of agricultural cooperatives in agro-products quality and safety control. *Issues in Agricultural Economy*, 9, 61-64.

Wang, Z., Mao, Y. & Gale, F. (2008). Chinese consumer demand for food safety attributes in milk products. *Food Policy*, 33(1), 27-36.

Wang, Z., Yuan, H. & Gale, F. (2009). Costs of adopting a hazard analysis critical control point system: Case study of a Chinese poultry processing firm. *Review of Agricultural Economics*, 31(3), 574-588.

Wei, L.& Lu, G. (2004). Functions of farmer specialized cooperatives on agro-products quality control: Case studies of several farmer specialized cooperatives in Zhejiang University. *Chinese Rural Economy*, 2, 36- 41.

Zhou, J. (2005). Study on vegetable quality and safety regulation: A case study of Zhejiang Province. *China Agricultural Press*, Beijing, China.

Implementation of Food Safety and Quality Standards: A Case Study of the Vegetable Processing Industry in Zhejiang, China

4.1 Introduction

China's vegetable industry has grown rapidly over the last decades. The consumption and production of vegetables has more than doubled. The per-capita consumption of vegetables had reached 270.49 kg in 2003, which was much higher than the world average per-capita consumption of 94.45 kg (Statistic of the World, 2008). Currently China, following Greece, has the second highest annual per-capita consumption of vegetables in the world. In addition, China has become a major exporting country of vegetables with an approximate value of $4.5 billion, an amount which accounts for 7% of world vegetable exports (FAO, 2004). These results show that China's vegetable industry is a very important agricultural sector domestically and is playing an increasingly important role in the world vegetable export market. The rapid growth of the vegetable sector has greatly contributed to China's domestic employment and farmer's income as well as poverty reduction due to the labor intensive nature of vegetable production. Vegetable production, without exception, in Zhejiang Province makes it one of China's leading areas in domestic and export markets. According to Zhejiang Agricultural Statistics, the production value of vegetables accounted for about 30% of the value of agricultural field-crop production in 2006, 40% of farm income came from vegetable production which absorbed approximately half of the province's agricultural labor force.

The rapid growth of the domestic and export vegetable markets, however, coupled with the poorly regulated food safety controls in China have led to a situation where vegetable growers rely heavily on agricultural chemicals (pesticides, fungicides, etc.) to control pest and disease problems to increase yield. As a result, agricultural chemical residues in vegetables are among the most common causes of food poisoning, which includes both acute and long term consequences as well as the effects of underlying diseases (Deng *et al.*, 2003; Li, 2002). Therefore, use of pesticides entails health risks for farmers, consumers and the environment. Participation in global trade means that China's vegetable industry faces international rules and increased scrutiny worldwide. Developed countries like Japan and Europe have strict food safety and quality standards and have passed them on to their suppliers in developing countries (Dolan and Humphrey, 2000). International and domestic requirements for quality and safety management in food industry have been tightened.

World trade in vegetables, as well as domestic demand for them, is likely to continue to grow. To keep long-term comparative advantage in this sector, the improvement of food safety is a top priority in China from both the domestic and the trades point of view. In response to the recent cases of food safety threats in China, especially related to pesticide residues, the Chinese government has placed greater emphasis on food safety improvement in the food supply chain. The government has taken major efforts by investing about

$5,000,000 annually to guarantee food quality/safety and improve the supply chain management that was started in 1999. By 2006, the government had also established a food safety system and developed more than 1800 standards and requirements of food quality/safety control on agricultural products. Among them, 634 standards are mandatory nationally. China has been able to achieve much in a relatively short time. For instance, the total numbers of firms and products that got the "Green Product" certification have increased 29.5% and 36% respectively between 2001 and 2007. However, the present control system still has major weaknesses. In 2007, the supply of agricultural products — cereals and oil, vegetable, fruit, livestock and poultry, aquatic products, tea, and other produces, with pollution-free certifications reached 206 million tons, accounting for about 10% of the same kind of produces. The output of products with green certification and organic certification are around 830 million tons and 1.955 million tons covered 4% and only 0.001%, respectively. The accumulative total of the agricultural products certification rate was about 14% (Ministry of Agriculture, 2008).

In recent years, the vegetable processing industry has increased in geographic specialization in China. Most vegetable processing firms are located in the provinces of Shangdong, Fujian, Zhejiang, Xinjiang and Guangdong. Zhejiang Province ranks third in the export value of vegetables. Compared with vegetable processing firms in other provinces, the firms in Zhejiang Province are more export-oriented, therefore these firms have a relatively higher food safety level than their counterparts in most other provinces. For example, the number of firms that have "green" certifications ranked second in 2006 and 2007 in China (Green Food, 2010). The vegetable processing firms in Zhejiang Province are more export oriented, closely connected to the international market, and thus more affected by the Chinese food quality and safety issues in the international context compared with firms from other regions in China. This chapter aims to explore the incentives that influence China's vegetable processing firms' decisions to adopt food quality/safety standards. To date, there are no formal studies that systematically investigated how food processing firms comply with the wide range of food safety and quality standards in developing countries. To fill this void, this chapter explores the incentives that influence establishment decisions with regard to food quality/safety controls in the Chinese vegetable industry. More precisely, we investigate the relationship between the degree of implementation of food safety/quality standards and the firm's internal and external factors for adopting these standards. We conduct the analysis in two steps: 1) to identify the major factors that affect firms' food safety standards' "adoption status", that is, whether or not a firm adopted any of the standards; 2) to determine what influenced the "adoption intensity", that is, the number of standards a firm adopted. Finally the results are expected to provide policy implications for food safety regulation and promote adoption of standards in the Chinese vegetable processing sector.

The rest of the chapter is organized as follows: Section 4.2 presents the food safety and quality system in China, Section 4.3 thoroughly reviews the literature about firms' incentives to adopt food safety standards, Section 4.4 presents the data collection; Section 4.5 presents the empirical logistic models, estimation results of the models and discussions; and this is followed by conclusion and policy implications in the last section.

4.2 Food Safety and Quality System in China

Food safety and quality standards not only improve food safety but also serve as a mechanism of information communication between buyers and sellers, thus reducing the buyer's uncertainty of a product's attributes by providing information about the seller's production process. According to the increased concerns on food safety and quality, the Chinese government has developed new food legislation and food quality/safety requirements for production and handling processes. Consequently, a series of standards becomes increasingly popular, which can be categorized into two groups: product-oriented standards and processing-oriented standards. The information about food safety and quality standards listed in Table 4.1 presents us with a general view of the Chinese main food safety standards and differences from other countries.

Table 4.1 Description of food safety and quality standards

Name	Category	Standards	Certification Agency	Definition
Free-pollution product	Voluntary (Certification) Mandatory (Standards)	National and Local Standards	Chinese Agriculture Dep.	Products that can contain limited artificial fertilizers and chemicals
Green product	Voluntary	National and CAC Standards	China Green Food Development Center	Product that is energy or water efficient; use healthy, non-toxic materials; made from recycled or renewable sources, make current products you use more efficient or more durable
Organic product	Voluntary	National Standards (referring IFOAM)	Chinese Environment Protection Agency	Product that is grown without the use of conventional pesticides, artificial fertilizers, human waste, or sewage sludge; processed without ionizing radiation or food additives
ISO 9000	Voluntary	ISO	China Bureau Quality and Technical Supervision	International standard that specifies the requirements for a food safety/quality management system
GMP	Voluntary (Certification) Mandatory (Standards)	National (referring CAC and U.S. GMP)	China Ministry of Health; China Ministry of Agriculture	Quality assurance which ensures that products are consistently produced and controlled to the quality standards appropriate to their intended use during the whole supply chain

(To be continued)

(Table 4.1)

GAP	Voluntary	National Standards (referring EUROGAP)	China National Regulatory Commission for Certification and Accreditation	Standards that are adopted in the critical production steps to ensure the consistency of food safety
HACCP	Mandatory (Frozen) Voluntary (others)	International HACCP	China National Regulatory Commission for Certification and Accreditation	Systematic preventive approach which identifies potential food safety hazards, so that key actions, known as Critical Control Points (CCP's) can be taken to reduce or eliminate the risk of the hazards being realized
QS	Mandatory (28 categories Processing products)	National Standards	Inspection and Quarantine of China	China Standards that guarantee the food has passed the necessary quality and safety tests

The product-oriented standards include pollution-free certification, "Green" certification and "Organic" certification. They aim at assuring consumers that the products are produced through meeting preconditions with regard to soil, water and environment. For instance, pesticide residue standards should meet specific production, harvesting and handling requirements like pesticide usage, limits on artificial additives, hygiene standards and residual pesticide levels. Some of these standards like pollution-free certification and Green certification are more tightly controlled than requirements in international markets. Economic theory indicates that product standards are often cheaper to implement than process standards, as product standards have more flexibility for firms to choose the least costly production methods that meet the standards (Unnevehr and Jensen, 1999; MacDonald and Crutchfield, 1996). Information about product-oriented standards flows through to final consumers in the form of third party verification, thereby assuring consumers that the product is safe.

In contrast to product-oriented standards, processing-oriented standards are designed to assure customers that products are produced and/or handled following specific practices to maintain consistent quality/safety. Processing-oriented standards adopted in China usually include ISO 9000, GAP, HACCP, GMP and QS. Some of those standards are usually recognized internationally for controlling food quality/safety. HACCP, for example, is a detailed protocol of requirements for production at the processing level with tracking, tracing and certification requirements. Through the means of regulations, the government requires food processing firms to implement mandatory HACCP for frozen vegetables in order to assure food safety in foreign markets.

4.3 Theoretical Framework

Knowledge of the factors that influence adoption of food quality/safety standards is important in efforts by food processing establishments to adopt and incorporate such standards in the product handling and distribution. By knowing the relative importance of these incentives, it is possible to relate this information to the propensity to adopt safety and quality standards. The goal of this section is to describe different factors influencing the firms' needs and capacities to adopt food quality/safety standards in the context of the Chinese vegetable processing sector. A number of studies have focused on the empirical perspective of quality/safety control in the food sector.

Many studies have indicated that the regulatory environment and involvement in export markets provided incentives for firms to adopt appropriate food quality/safety standards so as to meet required legal obligations. For example, Herath *et al.* (2007) and Hobbs *et al.* (2002) proposed that market orientation played a positive role in adoption of food safety control in the Canadian Food Processing Sector. Henson and Caswell (1999) raised the issue that regulatory responses to ensure safer food were important determinants for firms to improve safety control.

The private incentives for adopting food safety and quality standards can be internally or externally motivated. Some studies have suggested that firms' decisions on safety control were the result of externally driven reasons. For example, firms need to meet safety and quality demands of consumers and downstream suppliers. Ollinger *et al.* (2004) found that the food safety requirements by downstream suppliers were positively correlated to the probability of adopting food quality standards. Lindgreen and Hingley (2003) investigated the importance of food safety requirements by large meat retailers for determining incentives for firms to implement food quality/safety standards in the United Kingdom.

Still other studies indicated that internal incentives are also important for a firm to be certified. When a supplier adopts a particular quality standard, the expectation is that the action will result in improved market access and higher returns. At the same time, adopting a standard may improve internal operational efficiency because of the standard's model and the need to document the production process. According to Zaibet and Bredahl (1997), increasing a firm's efficiency played a role in firms implementing the ISO standard, because it created motivation for cost reduction, which generated a need for food quality/safety control.

Adoption of various standards can play a role in mitigating transaction costs in the supply chain. A product's safety and quality attributes may not be directly observable and information about these attributes is not easy to acquire. Contracting is the most widely used form of vertical coordination in the agricultural industry. The contractual definition of quality focuses on the transactions between the business partners. During contract negotiations, buyers and sellers desire to protect themselves from the possible food risk and transaction costs arising from uncertainty regarding food quality attributes. Therefore, contract pressure forces firms to demonstrate the quality/safety of their products. Product liability, or other aspects of food safety regulations, may require a firm to trace its products through the production process and to identify potential sources of contamination. Transaction cost reduction is of great interest to standards adoption research. Some previous studies have examined the concept that reducing the transaction cost would also be an internal incentive to adopt food quality/safety standards because they tried to mitigate the costs associated with recall and disposal of contaminated food, as well as loss of market share (Caswell and Hooker, 1998; Holleran *et al.*, 1999; Thomsen and McKensie, 2001).

Certain incentives to adopt quality/safety standards may be common to all firms in the vegetable industry, whereas the extent of adoption will be firm specific and/or market specific. For example, the size of firms may influence the motivation to implement those standards because the market returns of adoption are different. Many studies have examined whether the adoption motivation is influenced by firm-specific and/or market-specific characteristics (MacDonald and Crutchfield, 1996; Henson and Holt, 2000; Henson and Northen, 1998; Krishantha *et al.*, 2006). However, most studies were confined to a single standard. Conclusions from these studies may not apply to other standards. Unlike these studies that focus on a single standard, our study investigates the use of multiple standards.

4.4 The Survey and the Data

We use data from a survey taken during October 2006 to May 2007 that was conducted with vegetable processing establishments in Zhejiang Province. There are around 800 vegetable processing firms in Zhejiang Province. The Zhejiang Department of Agriculture maintains a list of vegetable processing "flagship enterprises" that covers most (more than 90% of vegetable sales) of the vegetable processing firms in Zhejiang. We randomly selected our sample from this list. The sample was stratified across ten regions covering almost the whole of Zhejiang Province[1]. Before the general survey, preliminary interviews were carried out twice in two selected sites—Jinhua and Ningbo areas[2]. In each area, twenty firms were randomly selected from the list of enterprises provided by the Agricultural Department of Zhejiang Province. The questionnaire was modified and polished based on feedback from the preliminary surveys and professional input and was then sent to firms across the province.

The survey was carried out either by on-site visits or by mail. For the personal interview, the firm's managers were asked about adoption of standards, the characteristics of their firms, orientation of markets, resources allocated to quality/safety control, and about their motivations to adopt the standards. For the mailing method, a letter was distributed to the firm's managers explaining the purpose of the survey and requesting they mail the completed questionnaire back. A hotline was opened to help the managers better understand the survey questions. A total of 170 enterprises were selected to participate in the survey. The returned surveys included 41 invalid questionnaires with quite a large quantity of uncompleted information, which we removed

---

[1] Actually, Zhejiang Province is made up of 11 regions, that is, Hangzhou, Ningbo, Wenzhou, Jiaxing, Huzhou, Shaoxing, Jinhua, Quzhou, Taizhou, Lishui and Zhoushan. We excluded Zhoushan city as it is an island and we do not think there is any loss in generality based on this decision.

[2] According to Zhejiang Statistical Yearbook, Ningbo's vegetable production is the second highest in northwestern Zhejiang and Jinhua's production is the second highest in southwestern Zhejiang. We chose these two areas because the vegetable processing firms in these two areas are representative of those in Zhejiang province in terms of technological development level, firm sizes as well as other characteristics.

from the results. Finally, 139 questionnaires were accepted as valid. Eight standards were investigated in this analysis: Pollution — free certification, Green certification, Organic certification, GAP, GMP, HACCP, ISO 9000 and QS. The 139 respondents could be classified by product types and food quality/safety standards (Table 4.2).

Table 4.2 Summary of survey

Stratification variable	Number of firms	Ratio (%)
Standards:		
Green	33	24
Pollution-free	66	47
Organic	6	4
ISO	66	47
GMP	8	6
GAP	1	1
HACCP	49	35
QS	50	36
Product types:		
Cut fresh	66	47
Frozen	31	22
Can and juice	20	14
Preserved	33	24
Dry	29	21

For process-oriented standards, firms implemented ISO 9000, QS and HACCP. Since vegetable supply chains now become more complex, more than several standards may be implemented for a firm to decrease the possibility of potential points of contamination from farm to table. Since the improving market access and increasing competitiveness in the long run have been regarded as another important incentive for firms to adopt multiple standards, the number of certifications that a firm obtained can be looked at as the magnitude of the level of incentives that the firm has. What we should mention here is that multiple standards may incur a higher cost of compliance and block regional trade. Therefore, greater standards harmonization would be encouraged in agricultural industries in China. The respondents can be divided into the following six strata according to the number of standards implemented, value 0 represented no certification, value 6 represented six standards that firms certified or adopted from the above eight standards. All stratas of standards adopted by firms are summarized in Fig. 4.1.

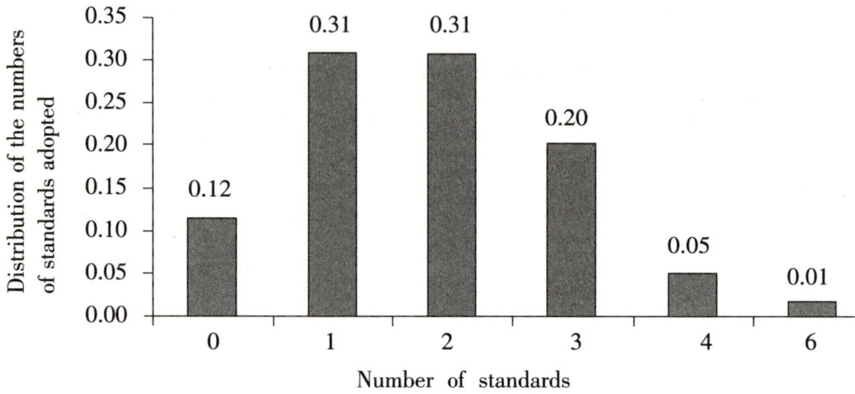

Fig. 4.1   **Distribution of standard numbers**

4.5 Empirical Analysis

The main drivers of a firm's propensity to adopt food safety/quality standards are the results that are realizable by adopting such standards. The adoption of such standards would impact on the economic outcomes of firms and such impacts would differ from one firm to another. The aim of the empirical analysis is to assess the interplay of factors that influence the decision and the intensity of firms to implement food quality/safety standards. An ordinal logistic model was constructed to illustrate probabilities of different levels of adoption intensity. This modeling approach is reasonable because the strategy of improving food safety is defined by a set of specific actions on a farm and within the supply chain where food quality/safety can be improved by means of control measures. Different types of standards can be adopted for a specific stage of a firm's operation. For instance, a firm may adopt "Green" standards on the farm level and implement "GMP" in the processing process. Thus, it is reasonable to believe that any of these standards will contribute to the level of food quality/safety.

An n-category ordinal logistic model which is used here is defined as

$$\log(p_i/1 - p_i) = \alpha_i + \beta'x, \; i = 1, 2, ..., n - 1$$

where p_i is the probability of being assigned to one of the categories, the logarithm of odds; x is a vector of independent variables. They are the driving forces of a firm's incentive to adopt quality/safety standards. β which is a vector of logistic coefficients. The intercepts vary between categories and satisfy the constraints $\alpha_1 \leq \alpha_2 \leq \, ... \, \leq \alpha_{n-1}$. It is assumed that the data are categorized independent of each other. Using the ordinal logistic setting, it is possible to estimate the relative odds of being in each category for firms which have a particular characteristic to those which do not after taking into account the effect of all other explanatory variables. The logistic coefficients represent the estimated increase of probabilities in each category of adoption intensity in the particular characteristics.

We estimate the probability of adopting a quality/safety standard. Our model draws upon the methods of Hassan *et al.* (2006). Firstly, we use a binomial logistic model to identify the factors that differentiate between adopters. A dichotomous variable takes value 0 for standards "less" adopters (zero or one standard)[3] and 1 for "more" adopters (two or more standards). Secondly, it is also important to understand to what extent a firm would implement food safety and quality standards. Ordered logistic analysis was then used to identify the difference between high-degree adopters and low-degree adopters. The adopting magnitude can be measured by a three-category scale ranging from "Low" to "Medium" and "High". Respondents that stated that no standards were implemented were classified as Low degree. Respondents that stated that one or two standards were implemented were classified as Medium degree. Respondents with three or more standards were classified as High degree. Therefore, ordered and binomial logistic models are specified as shown in Table 4.3.

---

[3] Agricultural standards system has gradually improved in Zhejiang Province; the ratio of pollution-free product standards covered around 81.3% at the end of the 10th five-year plan, and had reached 95% in end of 2009 (Source: www.zjbts.gov.cn). Another certification-QS is not mandatory standard but a basic one to enter market for food firms in Zhejiang Province. According to our investigation, most firms having either pollution-free certification or QS certification reached to a high ratio.

Table 4.3 Description of dependent variables

Dependent variable	Total adopted standards Y		Estimation method
Intensity of adoption	$Y=0$ (0)	Low	Ordered logistic
	$Y=1$ (1-2)	Medium	
	$Y=2$ (3 or more)	High	
Intensity of adoption	$Y=0$ (0, 1)	Less	Binomial logistic
	$Y=1$ (2 or more)	More	

A number of individual characteristics are expected to determinate the firm's activity on its adoption behavior. We hypothesise the following attributes to incentive firms' adoption motivations: (1) firm size; (2) brand; (3) E-commerce; (4) training frequency; (5) traceability; (6) expected premium; (7) export market; (8) supermarket; (9) government impacts. Detailed explanations on the variables' definition in the models are shown in Table 4.4, which also shows descriptive statistics.

Table 4.4 Definitions of independent variables

Independent variables	Description	Base	Mean
Firm size	>200 employees =1, otherwise = 0	<200 employees	0.30
	Or		
	100−199 employees =1, otherwise = 0	<100 employees	0.20
	200−499 employees = 1, otherwise = 0		0.24
	>500 employees = 1, otherwise = 0		0.08
Brand	With brand = 1, otherwise = 0	No brand	0.86
E-commerce	With E-commerce = 1, otherwise = 0	No E-commerce	0.54
Training frequency	Train times (year)		2.32
Traceability	With traceability =1, otherwise = 0	No traceability	0.16
Expected premium	With premium =1, otherwise = 0	No premium	0.59
Export market	Export market = 1, otherwise = 0	Wholesale/wet	0.21
Supermarket	Supermarket = 1, otherwise = 0	and other market	0.18
Government impacts	With impacts =1, otherwise = 0	No impact	0.49

The size of the firm

Large firms are expected to have greater access to financial resources, benefiting from economics of scale, having a relatively diverse workforce in terms of skills and being able to spread the fixed cost of adoption. They also have greater negotiating power over suppliers. In this study, firms are divided into four categories for the ordered model according to the number of employees: less than 100; from 100 to 199; from 200 to 499, and over 500. For the binomial model, firms are divided into two categories: less than 200 employees and over 200 employees.

Main market served

The pressure from downstream suppliers is one of the important factors that determine the propensity of standards adoption. With regard to the main markets served, we evaluate how much the presence of a processing firm on different markets influences its adoption behavior. Three types of markets are considered here: export market, supermarket, other markets which include wholesale market, wet/street market, restaurant and other institutions. Processing firms with higher export orientation are expected to more likely comply with quality/safety standards.

Expected premium

It is common in the literature that the hypothesis is that potential gains in terms of increases in market revenue allow the access to quality/safety control activities. A dummy variable was constructed scoring one if a firm has a high expected premium for being certificated, and zero otherwise.

Brand

Brand might be associated with reputation-related incentives for a firm to adopt food safety and quality standards. Firms with their own brands are expected to more likely implement a broader array of food quality/safety standards to compete in domestic and international markets. A dummy variable scoring one for a firm with its brand, and zero otherwise, were included in the model to capture the effects of brand on degree of adoption. Further, in existing literature, a rare account has been taken of the following factors that may influence the firms' incentive to adopt food safety quality controls.

E-Commerce

It is reasonable to expect that more innovative firms are more likely to implement food quality/safety standards, since those firms are more likely to explore ways to expand their market share (Herath *et al.*, 2007). In China, E-commerce can be looked at as an efficient way for firms to advertise themselves and obtain reputation gains. This makes more sense if we think that E-commerce can give an indication and the effects of product differentiation on standards adoption. It is a tool to efficiently measure the "differentiation" among firms that use mass media and web-based information rather than traditional marketing channels to promote their products.

Training frequency

A positive relationship between training activities and adoption degree was expected as firms with higher training frequency had a higher level of adoption degree. This suggests that firms are likely to be most responsive to the need for food safety when their laborers are trained by governmental agencies or professional institutions on the methods of enhancing food safety in the producing or handling process.

Government impacts

Exploring the relationship between the adoption of standards and government regulation actions would provide valuable information for policy makers regarding the role of regulation. In China, the government plays important roles in educating agricultural enterprises on food safety controls, providing technology and marketing services, strengthening food quality/safety legislation and enforcement of production. Thus, incentive-based approaches can help policy makers to understand the impacts of regulation on a firm's behavior and guild their efforts on interventions effectively.

According to Herath *et al.* (2007), a firm's characteristics are multidimensional and interactive. For example, the degree of export orientation might relate to its incentives for adopting food safety controls, such as reducing transaction costs, meeting legal requirements and responding to customer pressure. Thus, other variables were not included in the model due to multicollinearity problems. Also, the t value of the coefficient of the omitted variables is not significant, indicating that omitting these variables does not significantly change the model. The mean of the variables is listed in Table 4.4.

The regression results from binomial logistic and ordinal logistic regressions for the adoption of food quality/safety standards are presented in Table 4.5. The results indicate that the firm's characteristics are closely associated with the intensity of adoption. All the estimated coefficients exhibit plausible signs. In addition, most of the coefficients are statistically significant. At the same time, the ordered and binomial models produce similar patterns but with different magnitudes. The F-value of the adjusted Wald test for goodness of fit is significant at the 1% level for both models.

Table 4.5 Results of logistic regression model

Variables	Ordered logistic coefficient	Binomial logistic coefficient
Intercept 2	−5.241[*]	
Intercept 1	−1.137[***]	−3.848[*]
Firm size		
100–199 employees	0.990[***]	
200–499 employees	1.578[*]	
>500 employees	2.439[*]	
>200 employees	-	2.018[**]
Brand	1.245[**]	-
E-Commerce	0.760[***]	1.678[*]
Training frequency	0.455[**]	0.676[**]
Traceability	1.521	2.732[***]
Expected premium	0.882[***]	1.220[**]
Export market	0.667[***]	0.777[***]
Supermarket	0.015	0.757
Government impacts	0.086	0.729[***]
F-value	39.2[*]	42.3[*]

Note: ***, **, and * denote confidence level of 99%, 95%, and 90%, respectively.

The results indicate that a firm's size plays an important role in adopting food quality/safety standards. Particularly, the likelihood of standards adoption increases from the smallest to the largest establishments, which means larger firms are more likely to implement food quality/safety standards than smaller firms. This fact is consistent with the evidence in previous literature.

The results also show that the export market has a positive and significant coefficient with adoption standards in both the binomial and ordered model. The supermarket, however, is not significant in the adoption activities in both models, which is contradictory to our expectation. The product quality/safety level in supermarkets in China is significantly higher than that in farm markets. The insignificance of the supermarket coefficient in the model is probably because product share for supermarkets is far lower than the share for farm markets. In our samples, 63% of firms mainly served farm markets, only 14% of firms mainly served the supermarkets.

Not surprisingly, firms with higher expectations for a premium are more likely to invest in food safety and quality control. It is easy to understand and confirm the strong link between expected returns and adopting behavior. The variable is also the main contributor in both models to the discrimination between categories of the different intensity of adoption.

Firms with a brand tend to adopt a broader array of standards. However, a brand is not included in the binomial model. This is because the processing firms with brands are mostly firms of large size. When firms are not classified in several groups like the ordinal model, the variable of firm size and brand have a high correlation, which would cause the multicollinearity problem. The coefficient between e-commerce and adoption decision is significant and positive in both models, e-commerce focuses on network marketing. To our surprise, a large proportion of firms issue their product information and brand image via e-commerce according to our survey. They place great emphases on adopting food quality/safety standards to improve safety control in consideration of their brand image away from food safety events.

An increase in training frequency increases the probability of implementing more standards. The finding indicates that firms are more able to invest in food quality/safety control if they are trained. It has important implications for policy makers and regulating institutions since it shows that food quality/safety can be improved by providing training services.

The traceability variable indicates that the adoption decision among vegetable processing firms is associated with higher probability of being traced back in case of food accidents. According to the results, traceability increases the probability of adoption in the binomial model. However, traceability becomes insignificant in the ordinal model. This indicates that it is less important for motivating the firm to implement a broader range of standards.

The government regulation brings different impacts to the two models. In the binomial model, we find that the government regulation plays a significant role. It is, however, insignificant in the ordered model, which confirms the influence of government cannot contribute to a higher degree of adoption probability. The results imply that governments and administrating institutions should improve efficiency of food quality/safety interventions in order to enhance the functioning of markets. According to this survey, targeted supports in the form of financing and certification information are mostly favored by the interviewed firm managers.

The influence of an independent variable is calculated by comparing the probability when the variable takes a specific value with the probability of taking another specific value. The marginal probability is estimated to illustrate the difference between the mean probability values of different variables. The values of marginal probabilities are reported for both the ordered model and the binary model (Table 4.6). Similar to the finding in Hassan *et al.* (2006), establishment size has a great influence on the adoption decision of food quality/safety standards. Furthermore, the marginal contribution of probability increases with the increase in firm size in the ordered model. Traceability, the ability to trace the origin of raw materials throughout the supply chain, has a pronounced effect on the probability of adoption (equal to 29%), relative to other characteristics in the binary model. It is easy to understand since a firm has higher likelihood of being identified through traceback and assigned responsibility of producing unsafe food with a higher traceability rate. However, traceability has a much lower contribution to standards implementation in the ordered model compared with that in the binary model. This result suggests that traceability has a less differentiating effect on a firm's decisions in the ordered model. For the other variables, the marginal probabilities in the two models only have small differences. Specifically, we can find expected premium and e-commerce having a stronger influence than the level of export orientation, training frequency and government impact on the adoption of quality/safety standards in both models. The results suggest that the market-based private incentives have a greater impact on a firm's adopting decision than government regulation. Furthermore, the contribution of the export market is only 9% in the binary model. This was closely related with the market share. In our survey data, only 22.3% of the total numbers of enterprises serve foreign markets. The contribution of expected premium to the adoption probability is 13%, while the marginal probability of government impact is only about 7% in the binary model.

Table 4.6 Marginal probability

Variables	Marginal probability (%)			
	Low	Medium (Ordered logistic)	High	More (Binomial logistic)
Firm size				
100–199 employees	– 5	– 13	18	–
200–499 employees	– 7	– 26	33	–
>500 employees	– 7	– 47	54	–
>200 employees	–	–	–	21
Brand	– 12	– 5	17	–
E-commerce	– 5	– 8	13	18
Training frequency	– 3	– 5	8	7
Traceability	– 2	– 5	7	29
Expected premium	– 6	– 9	15	13
Export market	– 1	– 3	4	9
Supermarket	– 1	– 1	2	8
Government impacts	0	– 1	1	7

The results of logistic analysis indicate that the primary market served significantly influenced the firm's incentive for adopting food quality/safety standards. The following analysis focused on influences of downstream supports on a firm's decisions. The downstream supports are named as supporting activities that downstream suppliers provided to help the processing firms to comply with specific food quality/safety requirements so as to be certified. Typical support in the vegetable processing sector can be summarized in five categories: (1) raw materials support; (2) technical support; (3) fund support; (4) training support; and (5) certificate support. Survey results show firms get more support in export markets compared with other markets: 41.9% of the importers provided raw materials support to processing firms, 56.5% of importers provided technical support, 19.4%, 43.5% and 41.9% of importers provide funding support, training support and certification support, respectively (Table 4.7 and Fig. 4.2).

Table 4.7 Downstream support

Markets	Raw materials	Technical	Fund	Training	Certification	Chi-square (*p*-value)
Export market	26 (41.9%)	35 (56.5%)	12 (19.4%)	27 (43.5%)	26 (41.9%)	-
Supermarket	17 (16.8%)	22 (21.8%)	19 (18.8%)	39 (38.6%)	33 (32.7%)	9.52 (0.02)
Wholesale market	28 (18.2%)	26 (16.9%)	23 (14.9%)	19 (12.3%)	19 (12.3%)	-

The bivariate analysis is used to test whether or not the adoption decision and support from different served markets are independent. The support includes every type mentioned above.

The results suggest that, for the adopters, the likelihood of receiving support in export markets, supermarkets, processing markets and wholesale markets decreases from the largest to the smallest. Chi-square test (Table 4.7) showed the proportional odds assumption was accepted when applied to the markets. The explanation for this is that the adopters value "support from served markets" as important factors when deciding whether to adopt food quality/safety standards or not. In contrast, "Less" adopters did not seem to care as much about export markets and supermarkets as did the adopters.

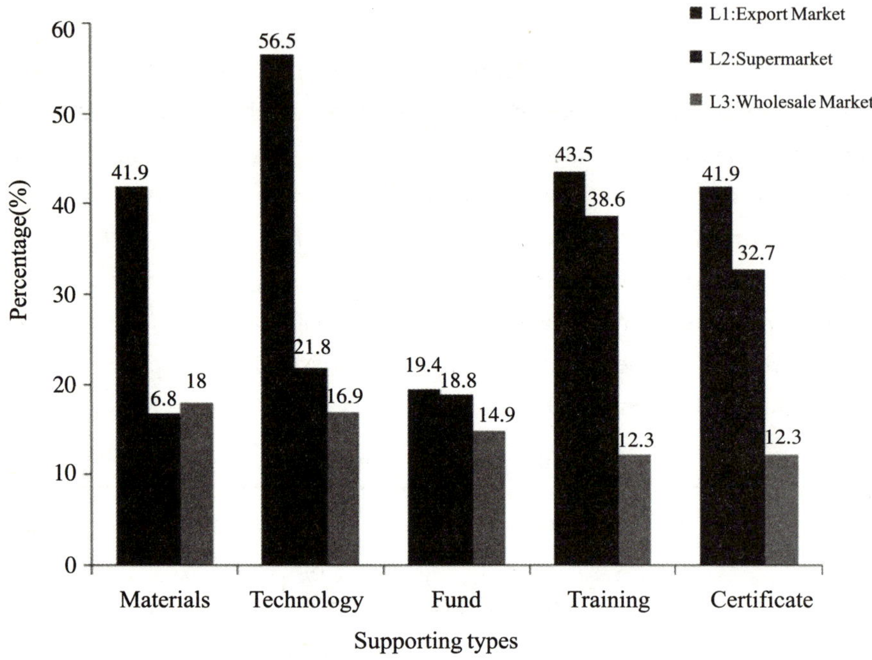

Fig. 4.2 Downstrcam supports

4.6 Conclusions and Policy Recommendations

China's vegetable industry is a rapidly growing sector. Over the last decade, great attention has been paid to food quality/safety control due to the increase in pesticide residue related diseases, in particular in China. To meet the demands of food quality/safety standards and the requirements of government regulations, firms in the Chinese vegetable industry have adopted a number of standards to guarantee safe vegetable products.

This study empirically investigates the factors that influence adoption behavior and adoption magnitude of food quality/safety standards in the Chinese vegetable processing sector. Based on the previous studies, nine factors such as firm size, export market, supermarket, expected premium, brand, e-commerce, training frequency, traceability and government impact are expected to incentivize and impact agricultural firms' propensity and degree of adoption decisions. The results suggest that the significance and magnitude of the coefficients associated with the independent variables differ between binomial and ordered logistic models. A firm's standard adoption intensity is firstly affected by firm size. Referring to target markets, the export market-relative to supermarkets, wholesale markets, street markets and other traditional markets, has stricter quality requirements, thus it has a stricter control for the vegetable processing procedure. But export markets also provide quality control support, which makes it easier for firms to implement the quality and safety standards. Similarly, a positive relationship can be found between the magnitude of adoption and training activities. The findings indicate that government plays an important role in the firms' adoption of vegetable quality and safety standards, while the major driving force comes from the market, that is, whether or not "good quality" can cause "good price". When individual firms choose whether to implement quality standards or not, they weigh their private benefit and cost. In a word, a processing firm of large size, own brand, and e-commerce is more likely to adopt those kinds of quality standards.

As the Chinese vegetable processing industry is still young but is developing at a fast speed, it is important to set up a mechanism that combines market driven forces and governmental policies. Based on the above conclusions, we identified the following needs and recommendations for improved policy and standards to support safety and quality in the marketing system.

Stricter regulations such as a market access system are required to incentivize firms to universally implement food quality/safety standards. In some cases, agricultural firms that produce products with inferior quality force legitimate producers out of the market by continuously cutting cost. Hence, a firm's high propensity and degree of adoption decisions can assure that only safe products certified by quality/safety standards that meet certain standards can enter the markets.

Based on the precept that "good quality" can bring "good price", it is important to educate the public about the product standards and certifications. Consumers' recognition of the certification can lead them to pay a higher premium for certified products, which can provide a market driven force for firms to adopt the standards.

According to our qualitative analyses, large scale firms are more likely to adopt food safety standards whereas the scale of the Chinese vegetable processing firms is relatively small and lag behind technologically. Additional resources are needed to support research and development of food safety control techniques, and to facilitate the firm's cost of food safety control. What's more, the research should also focus on technologies and methods that can be used by smaller firms. These technologies can be supported by expanded training that may act as a bridge to connect firms and higher educational institutions.

Finally, additional support is required to encourage modern retail food marketing networks like supermarkets. Supermarkets with a higher level of food quality/safety standards ensure more safe food being supplied to consumers. However, firms, from our investigation, have a smaller market share in supermarkets because it's difficult to meet the limit of strict market access now, as well as Chinese consumers' traditional custom to buy fresh and processed vegetables in farm markets. Despite of this, such an approach would be widely accepted for its significant economic and social benefits in the long run, especially relative to its improved food safety and quality.

References

Caswell, J. & Hooker, N. (1998). How quality management metasystems are affecting the food industry. *Review of Agricultural Economics*, 20, 547-557.

Deng, L., Qu, H., Huang, R., Yang, Y., Zhen X. & Wang, H. (2003). Survey of food poisoning by organosphorus pesticide at an employee refectory. *Practical Preventive Medicine*, 10(5), 766-767.

Dolan, C. & Humphrey, J. (2000). Governance and trade in fresh vegetables: Impact of UK supermarkets on the African horticultural industry. *Journal of Development Studies*, 37, 147-177.

FAO (2004). The Market for Non-Traditional Agricultural Exports. Rome.

Green Food (2010). Http://www.greenfood.org.cn/sites/MainSite/List_2_2453.html.

Hassan, Z., Green, R. & Harath, D. (2006). An empirical analysis of the adoption of food safety and quality practices in the Canadian food processing industry. Essays in Honor of Stanley R. Johnson. Article 18.

Henson, S. & Holt, G. (2000). Exploring incentives for the adoption of food safety controls: HACCP implementation in the UK dairy sector. *Review of Agricultural Economics*, 22, 407-420.

Henson, S. & Northen, J. (1998). Economic determinants of food safety controls in supply of retailer own-branded products in United Kingdom. *Agribusiness*, 14(2), 113-126.

Henson, S. & Caswell, J.A. (1999). Food safety regulation: An overview of contemporary issues. *Food Policy*, 24, 589-603.

Herath D., Hassan, Z. & Henson, S. (2007). Adoption of food safety and quality controls: Do firm characteristics matter? Evidence from the Canadian food processing sector. *Canadian Journal of Agricultural Economics*, 55(3), 299-314.

Hobbs, J.E., Fearne, A. & Spriggs, J. (2002). Incentive structures for food safety and auality assurance: an international comparison. *Food Control*, 13, 77-81.

Holleran, E., Bredahl, M.E. & Zaibet, L. (1999). Private incentives for adopting food safety and quality assurance. *Food Policy*, 24, 669-683.

Krishantha, U., Mudalige, J. & Henson, S. (2006). Economic incentives for firms to implement enhanced food safety controls: Case of the Canadian red meat and poultry processing sector. *Review of Agricultural Economics*, 28, 494-514.

Li, X. (2002). Analysis of food poisoning due to taking vegetable contaminated with organophosphorus pesticide. *China Tropical Medicine*, 2 (4), 519.

Lindgreen, A. & Hingley, M. (2003).The impact of food safety and animal welfare policies on supply chain management. *British Food Journal*, 105, 328-349.

MacDonald, J.M. & Crutchfield, S. (1996). Modeling the costs of food safety regulations. *American Journal of Agricultural Economics*, 78, 1285-1290.

Ministry of Agriculture (2008). China Agricultural Statistics Yearbook. Beijing: China Agricultural Press.

Ollinger, M.E., Moore, D. & Chandran, R. (2004). Meat and Poultry Plant's Food Safety Investments: Survey Findings. Technical Bulletin No. 1911, United States Department of Agriculture, Economics Research Service.

Statistic of the World (2008). Statistics Bureau and the Director-General for Policy Planning (Statistical Standards) & Statistical Research and Training Institute of Ministry of Internal Affairs and Communications of Japan. Beijing: Statistics Bureau Press.

Thomsen, M.R. & McKenzie, A.M. (2001). Market incentives for safe foods: An examination of shareholder losses from meat and poultry recalls. *American Journal of Agricultural Economics*, 83, 526-538.

Unnevehr, L.J. & Jensen, H.H. (1999). The economic implications of using HACCP as a food safety regulatory standard. *Food Policy*, 24(6), 625-635.

Zaibet, L. & Bredahl, M. (1997). Gains from ISO certification in the UK meat sector. *Agribusiness: An International Journal*, 13(4), 375-384.

Chapter

5

Adoption of HACCP System in the Chinese Food Industry: A Comparative Analysis

5.1 Introduction

China has witnessed a rapid increase in the export of food products during the 1990s because of low production costs. In recent years, however, China's exports have suffered a lot as they cannot always meet international food safety standards. The Hazard Analysis and Critical Control Point (HACCP) system was then introduced and extended by the government to enhance the safety of foods and to close the gap between Chinese and international food safety standards. The HACCP system is particularly recommended for export-oriented food enterprises as a sanitary standard in international trade. On the other hand, with the rapid growth of per capita income and living standards, food safety is of increasing concern for public health in China (Bai *et al.*, 2007a). As is argued by Bai *et al.* (2007b), domestic consumers are entitled to the same food safety standards as foreign consumers. Implementing the HACCP system is therefore of extreme importance, not only for export growth but also for the welfare of the domestic consumer. According to the China Statistical Yearbook of Certification and Accreditation (2005), however, although the number of enterprises that have adopted an HACCP system has increased from 2003 to 2005, its presence throughout the food industry as a whole has fallen from 23.1% to 21.9%. Fig. 5.1 gives the details of the changes from 2003 to 2005. The rate of implementation of the HACCP system in China is still very low, which is in sharp contrast to that in developed countries[1].

There has been a lack of research, especially empirical research, into the issue of HACCP system implementation in China. China's food safety assurance system was first introduced by Bai *et al.* (2007a) and divided China's food safety assurance system into a compulsory food safety administration system (e.g. Food Quality Safety Market Access System) and voluntary food safety consumer assurance systems (e.g. Green Food Certification system, Organic Food Certification system and HACCP system). Moreover, based on a survey of 27 food enterprises which had implemented the HACCP system, Bai *et al.* (2007b) reported on the characteristics of those food enterprises and incentives for them to implement the HACCP system. There is little available literature that either describes the types of Chinese food manufacturers who do not adopt the HACCP system or suggests methods to encourage them to do so. This study, therefore, differentiates itself from prior research by focusing on those food enterprises which have not adopted the HACCP system and reports the survey results by comparison with food enterprises that have already adopted the HACCP system.

---

[1] Based on a survey of the food industry in the Yorkshire and Humberside region in the UK regarding the implementation of the HACCP system, Panisello et al. (1999) reported that 72.6% of food companies had implemented the HACCP system. Also, Henson et al. (1999) found that 73.9% of survey respondents in the dairy industry in the UK claimed that they had a fully operational HACCP system in place. With regard to other developed countries, Unnervehr et al. (1999) affirmed that the European Union, the United States and Australia have mandatory programs to encourage the adoption of the HACCP system.

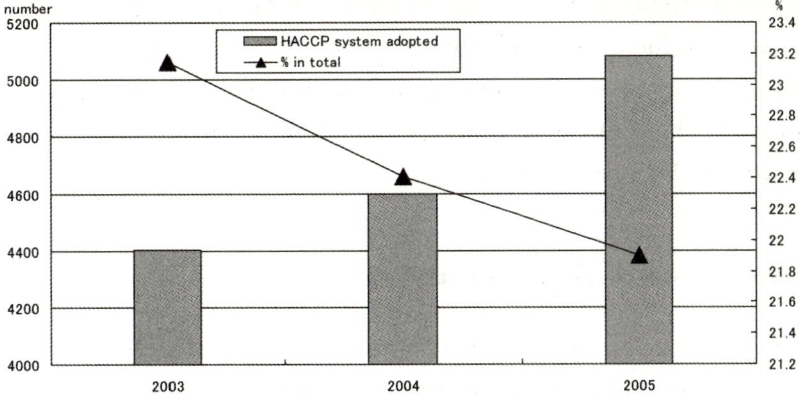

**Fig. 5.1   Number and percentage of food enterprises
that have adopted the HACCP system**

Data source: China Statistical Yearbook of Certification and Accreditation (2005)

5.2 Method

As is employed in other similar studies, a questionnaire survey was selected
for our study. Firstly, focus group interviews were conducted before the questionnaire was developed. We randomly selected eight food enterprise managers to form a focus group and interviewed them with regard to the implementation of the HACCP system in their enterprises. The information gathered
from the focus group interviews has been used as the source for the survey
questions.

In the second phase the questionnaire, made up of closed questions, was developed and divided into different themes. In part I we asked basic questions
about the food enterprises and their managers. Part II included questions related to the manager's knowledge of the HACCP system and whether other accreditations had also been implemented to secure food safety. Part III consisted of 12 questions related to motives and external factors that may affect
HACCP system adoption.

A mailing list of 160 food enterprises in Zhejiang Province was selected
from the list provided by Hangzhou Agricultural Technology Information Ser-

vice Center, Zhejiang Province based on their HACCP system implementation status (80 food enterprises have a fully operational HACCP system, the others have not)[2]. The questionnaires were mailed to the directors or managers of the food enterprises in February 2006. Completed questionnaires were received from February to May 2006.

5.3 Results and Discussions

A total of 132 questionnaires were returned. After eliminating those questionnaires with incomplete responses and those that were otherwise unusable, 117 usable respondents were obtained for an effective response rate of 97%. 51 of the 117 usable respondents were from enterprises that had not adopted the HACCP system and the remaining 66 were from enterprises that had a fully operational HACCP system in practice. We coded the survey and employed the Statistical Package for the Social Sciences (SPSS) to analyze the data.

5.3.1 Demographic Analysis

Demographic information related to the characteristics of the food enterprises and their managers is presented in Table 5.1. The main business of the food enterprises surveyed was vegetable and fruit processing or meat processing. The majority of survey respondents were private enterprises, followed by joint-venture foreign enterprises, collective enterprises, sole investment foreign enterprises, and finally state-owned enterprises. About 53% of the food enterprises without an HACCP system in practice were private enterprises, which is a higher percentage than those food enterprises with a fully operational HACCP system (39.1%). Compared with the respondents without an HACCP system in practice, the respondents with a fully operational HACCP system in practice had more employees, with 48 (76.2%) of them employing

---

[2] The proportion of food enterprises that have not implemented HACCP in our sample is not consistent with the China Statistical Yearbook of Certification and Accreditation (2005). The sample was controlled so that we could carry out a comparative analysis.

**Table 5.1   Demographic characteristics of the food enterprises
and their managers**

Characteristics		Frequency (valid %)	
		HACCP status	
		No	Yes
Food enterprises			
Main business	Vegetable and fruit processing	9 (17.6)	15 (23.8)
	Meat processing	18 (35.3)	18 (28.6)
	Others	24 (47.1)	30 (47.6)
Type of enterprises	State-owned enterprises	1 (2.0)	10 (15.6)
	Private enterprises	27 (52.9)	25 (39.1)
	Collective enterprises	9 (17.6)	8 (12.5)
	Joint-venture foreign enterprises	13 (25.5)	9 (14.1)
	Sole investment foreign enterprises	1 (2.0)	12 (18.8)
Number of employees	Less than 100	24 (47.1)	6 (9.5)
	From 100 to 500	21 (41.2)	9 (14.3)
	From 500 to 1000	6 (11.8)	24 (38.1)
	More than 1000	0	24 (38.1)
Market strategy	Foreign market oriented	27 (52.9)	45 (71.4)
	Domestic market oriented	24 (47.1)	18 (28.6)
Other accreditations implemented	GMPs	6 (11.8)	33 (50.0)
	SSOPs	6 (11.8)	36 (54.5)
	ISO 9000 series	21 (41.2)	48 (72.7)
Managers			
Age	Under 30 years	24 (47.1)	21 (33.3)
	From 31 to 40 years	15 (29.4)	24 (38.1)
	From 41 to 50 years	9 (17.6)	9 (14.3)
	51 years and over	3 (5.9)	9 (14.3)
Education level	Middle school or below	3 (5.9)	3 (4.5)
	High school	12 (23.5)	12 (18.2)
	College	24 (47.1)	21 (31.8)
	Postgraduate	12 (23.5)	30 (45.5)

more than 500 workers. Our result strongly agrees with that of Panisello *et al.* (1999), which reported that small businesses were less likely to have implemented an HACCP system than their larger counterparts in the UK. Although most respondents focused on both domestic and foreign markets, when they were asked to indicate their market strategy, 27 of those (52.9%) without an HACCP system in place indicated that they focused on the foreign market, while 71.4% of those with a fully operational HACCP system targeted the foreign market. We also asked respondents whether they had implemented any other non-compulsive accreditations. More than half of the respondents who had adopted the HACCP system claimed that they had also implemented other quality management systems such as Good Manufacturing Practices (GMPs), Sanitation Standard Operating Procedures (SSOPs) and International Standards Organization 9000 (ISO 9000) series to ensure food safety. For the respondents without an HACCP system in place, only 11.8% of their enterprises implemented GMPs or SSOPs and 21 (41.2%) enterprises employed the ISO 9000 series.

In this study, the average age of the managers was 35 years. The majority of managers (more than 70%) were under 40 years of age. Regarding the educational status of the managers, 36 managers (70.6%) of food enterprises without a fully operational HACCP system and 51 managers (77.3%) from food enterprises that had implemented the HACCP system had a college or higher main degree. The higher the education level of the managers, the more likely it was that their company had adopted the HACCP system. As for the educational status, the majority of survey respondents had a college or postgraduate degree. The survey sample indicates that on average managers of food enterprises with a fully operational HACCP system in place are better educated than those of food enterprises without an HACCP system in place.

5.3.2 Perceptions about the HACCP System

Table 5.2 presents a comparison between different food enterprise's perceptions of the HACCP system. Firstly, we asked the respondents about the losses suffered through food safety accidents. The results are very interesting and show that 22.7% of the food enterprises with a fully operational HACCP system in place suffered losses as a result of food safety accidents, while none of the food enterprises without an HACCP system in place suffered any large losses due to food safety accidents. This implies that to some extent Chinese food enterprises adopted the HACCP system maybe due to their losses caused by food safety accidents. Moreover, there were also 21 (31.8%) respondents whose companies had implemented an HACCP system who answered "never" to the question regarding problems with food safety, which is higher than (11.8%) the number of respondents without an HACCP system in place who gave the same answer, which suggests that although implementing the HACCP system is a burden for most food enterprises[3] (Taylor and Kane, 2005), it does ensure food safety.

Unlike non-HACCP-implemented respondents, the majority of implemented respondents had a good knowledge of the HACCP system. Sixty-three out of 66 respondents with a fully operational HACCP system in place indicated that they had an exact knowledge of the HACCP system, in contrast to managers of enterprises without an HACCP system in place that had little information regarding the HACCP system, which partially explains why their enterprises had not implemented it. Furthermore, we asked respondents whether the HACCP system was effective in controlling food quality, and as many as 15 (30%) respondents without an HACCP system in place answered negatively, which indicates that the doubtful attitude of the Chinese food industry regarding the effectiveness of the HACCP system hampers its implementation.

---

[3] Taylor and Kane (2005) reported that HACCP system implementation was a burden for food enterprises in the UK. This is especially true for a developing country like China. As GMPs/SSOPs, the pre-requisite programs for HACCP system implementation, were also introduced recently in the Chinese food industry, most food enterprises that want to implement the HACCP system have to implement HACCP together with GMPs/SSOPs. High costs mainly result from expenditure on GMPs/SSOPs improvements.

Table 5.2 Perceptions of the HACCP system

		HACCP status	
		No	Yes
Losses suffered from food safety accidents	Yes	0	15 (22.7%)
	Sometimes	45 (88.2%)	30 (45.5%)
	Never	6 (11.8%)	21 (31.8%)
Knowledge of HACCP	Yes, completely	9 (17.6%)	63 (95.5%)
	A little	39 (76.5%)	3 (4.5%)
	None	3 (5.9%)	0
HACCP system is effective for controlling food quality	Agree	27 (52.9%)	63 (95.5%)
	Somewhat agree	9 (17.6%)	3 (4.5%)
	Disagree	15 (29.4%)	0

5.3.3 Motives and External Factors to Encourage the Adoption of the HACCP System

In part III we listed 12 items related to motives and external factors affecting HACCP system implementation and asked respondents to rate their answers on a five-response Likert scale ranging from "very important" to "very unimportant"[4]. We calculated the mean of each item for both groups, and then performed an analysis of variance (ANOVA) to highlight any differences between the group means. The results of the statistical analysis are shown in Table 5.3.

The most important motive for survey respondents without an HACCP system in place was to improve product quality, but for those food enterprises which had already adopted the HACCP system, the motive was to lower the risk of compromising food safety. The ANOVA revealed that differences exist between both groups on whether implementing the HACCP system would lower the risk of compromising food safety, expand their foreign market or improve their profit margin. It is notable that food enterprises with a fully operational HACCP system in place are more likely to view implementation of the HACCP system as a way of improving their profit margins. This may imply that the benefits of implementing the HACCP system came from revenue

---

[4] To conduct the analysis, each item is assigned a score from 1 to 5, 1 stands for "very important" and 5 represents "very unimportant".

increases and cost savings due to reduced product losses resulting mainly from GMPs/SSOPs improvements, which covered high implementation costs. Furthermore, respondents were asked to rate the importance of four external factors that may affect their decision as to whether or not to adopt the HACCP system. The most important of these was the consumer's awareness of food safety, followed by extension and support from the government, compliance with the law and the recommendations of industry associations. On average, the external factors were more important to survey respondents who had not implemented the HACCP system than to those who had a fully operational HACCP system in place.

Table 5.3 Motives and external factors to affect the adoption of the HACCP system

	HACCP status	Mean	F test
Motives			
To improve product quality	No	1.41	0.115
	Yes	1.45	
To lower risk of compromising food safety	No	1.56	2.990*
	Yes	1.32	
To expand foreign markets	No	1.56	3.692*
	Yes	1.91	
To take a leadership position	No	1.76	1.425
	Yes	1.95	
To build a strong brand	No	2.00	0.000
	Yes	2.00	
To reduce waste	No	2.06	0.007
	Yes	2.05	
To improve profit margins	No	2.44	14.259***
	Yes	3.05	
To obtain other third party accreditations	No	2.47	1.857
	Yes	2.76	
External factors			
Consumer awareness of food safety	No	1.71	4.054**
	Yes	2.00	
Extension and support from the government	No	2.35	4.655**
	Yes	2.77	
Compliance with the law	No	2.63	0.003
	Yes	2.64	
Recommendation of industry associations	No	2.71	5.519**
	Yes	3.19	

Note: *, **, *** significant at 10%, 5%, and 1%, respectively.

5.4 Conclusions and Implications

From our comparative analysis, we can identify important policy implications to encourage the implementation of the HACCP system in the Chinese food industry. Based on the results, a profile of food enterprises most likely not to adopt an HACCP system are:

(1) Small and medium-sized enterprises (SMEs) with less than 500 employees. As is intensively discussed in Taylor and Kane (2005), the introduction of the HACCP system in SMEs is a burden because of the high cost of implementation. It is inappropriate to force SMEs to implement the HACCP system in China as they lack the financial resources and other relevant equipment (Bai *et al.*, 2007b). What the government can do, therefore, may be to select sample food enterprises and try to help them establish HACCP systems by providing adequate cost-benefit information, which would enable other SMEs to learn from the sample enterprises. In addition, it is necessary to set up province or city level HACCP resource centers to provide professional training and consultation services on HACCP system implementation.

(2) Suppliers of the domestic market. Food markets in developed countries are always characterized by strict food regulation systems and the majority of Chinese food enterprises have adopted the HACCP system to satisfy them. The low implementation rate of the HACCP system in the domestic market clearly indicates a lack of true commitment to food safety management. As is argued in Bai *et al.* (2007a), although the implementation of the Food Quality Safety Market Access System has been made compulsory to assure food quality, it is still inefficient. To some extent, setting up a strict food regulation system in China will not only improve Chinese consumers' welfare but will also encourage food enterprises to adopt the HACCP system.

(3) Food enterprises that have not implemented other quality management systems such as GMPs or SSOPs. To stimulate the adoption of the HACCP system by food enterprises in China, GMPs and SSOPs should be encouraged, as they not only serve as the basis for HACCP system implementation but also help to cut the cost of implementation. This is especially true for the

SMEs which are financially able to implement such systems.

(4) Food enterprises that have managers with relatively low education levels. As results of this study show that managers of food enterprises with a fully operational HACCP system in place are better educated than those of food enterprises without an HACCP system in place, raising the education level of the managers may be a feasible solution to increase the implementation rate of the HACCP system in the Chinese food industry.

Perceptions of the HACCP system directly relate to the motives for adopting the HACCP system. The findings of this study indicate that, compared with the food enterprises that have implemented the HACCP system, those that have not appear to have a limited perception of the HACCP system. Thus, a major policy implication of this study pertains to the training of managers with regard to the HACCP system in Chinese food enterprises. A good knowledge of the HACCP system and food safety management overall will undoubtedly help managers to make a decision as to whether or not to adopt it.

As the results of our comparative analysis show that food enterprises without an HACCP system in place are more sensitive to food safety awareness of consumers, it would indeed be wise of policy makers in China to popularize scientific knowledge concerning food safety and a basic knowledge of the HACCP system for the public. Once the public are aware of the effectiveness of the HACCP system to ensure food safety, they are likely to prefer food products produced by those companies which employ an HACCP system. This will in turn prompt food enterprises to adopt the HACCP system. Furthermore, budgetary priority should be given to the extension of the HACCP system because, according to the results, extension and support from the government is especially important to those food enterprises that have not adopted the HACCP system. In addition, due to the lack of GMPs/SSOPs, an educated workforce and necessary equipment in the Chinese food industry, implementation of the HACCP system is a financial burden for food enterprises. Special loan support or preferential taxes should be considered in order to stimulate HACCP system implementation in the Chinese food industry.

References

Bai, L., Ma, C., Gong, S. & Yang, Y. (2007a). Food safety assurance systems in China. *Food Control*, 18, 480-484.

Bai, L., Ma, C., Gong, S. & Yang, Y. (2007b). Implementation of HACCP system in China: A survey of food enterprises involved. *Food Control*, 18, 1108-1112.

Henson, S., Holt, G. & Northen, J. (1999). Costs and benefits of implementing HACCP in the UK dairy processing sector. *Food Control*, 10, 99-106.

Panisello, P.J., Quantick, P.C. & Knowles, M.J. (1999). Towards the implementation of HACCP: results of a UK regional survey. *Food Control*, 10, 87-98.

Taylor, E. & Kane, K. (2005). Reducing the burden of HACCP on SMEs. *Food Policy*, 16, 833-839.

Unnevehr, L.J., Miller, G.Y., Gómez, M.I. (1999). Ensuring food safety and quality in farm-level production: emerging lessons from the pork industry. *Am. J. Agric. Econ.*, 81(5), 1096-1101.

An Empirical Analysis of the Implementation of Vegetable Quality and Safety Traceability Systems Centering on Wholesale Markets

The reason why agricultural product quality and safety issues happen now and then rest with the market failure is due to the asymmetric information and non-traceable responsibility. This conclusion has been recognized first by the policy makers in the EU and then also in other developed countries (Banterle and Stranieri, 2008; Caswell, 1998; De Castro,2002; Fan and Jin, 2006; Golan, 2004; Golan *et al.*, 2000, 2003, 2004, 2005; Hall, 2010), and such policy makers attach great importance to the establishment of a food quality and safety traceability system from the perspective of policy management. What's more, in some European countries, food is not allowed to circulate in the market if its safety and quality responsibility is non-traceable (Hobbs,

2004; Hobbs *et al.*, 2007). Currently, in China, wholesale markets are a key link in the main channel of vegetable circulation. Under the circumstances that the existing level of organization in agriculture is relatively low, the degree of market standardization is not high, the credit system is not perfect, and the financial resources of government are limited, it is therefore regarded as an effective measure to trace the vegetable quality and safety responsibility from the wholesale stage to the stage of production and consumption, together with the implementation of a market access system and the advancement of consumers' awareness by requesting certificates and invoices. To further explore the implementation mechanism of the traceability system, this chapter investigates the status quo and difficulties of the implementation of a vegetable quality and safety traceability system centering on wholesale markets from the perspective of suppliers of vegetable wholesale markets and relevant government departments for quality and safety management.

In this study, the sample data is based on the field research conducted by the author and other associates of the research group in the wholesale markets of such cities as Hangzhou, Jiaxing, Ningbo, Wenzhou in Zhejiang Province, and Weifang, Jining in Shandong Province during July and August 2009. In each city, the author randomly selected one municipal wholesale market and three county-level wholesale markets as the research objects; the author carried out a questionnaire survey on the vegetable suppliers of each wholesale market for a period of two to three days, including about 50 questionnaires for each municipal wholesale market and 20 for each county-level wholesale market. The survey involved a total of 28 wholesale markets, getting back 410 valid questionnaires and 90 invalid questionnaires for relevant government regulatory authorities.

6.1 The Status Quo and the Reasons for the Implementation of a Traceability System in China's Agricultural Products Wholesale Markets

Agricultural products wholesale markets are the gathering places of agricultural logistics and information flow and are important stages of agricultural products quality and safety management. In order to ensure the quality and safety of agricultural products in the circulation, the relevant departments of the central government and of all provinces have been vigorously promoting the construction of agricultural products quality and safety traceability systems centering on wholesale markets and have made obvious progress in implementing such systems.

The fundamental condition for a vegetable traceability system can be assumed to be that when vegetables enter or leave a wholesale market, the quality and safety responsibility of such vegetables are traceable. According to this standard, on the whole, the percentage of traceable vegetables in China's vegetable wholesale markets is relatively low. This survey shows that when entering the market, traceable vegetables accounted for 45.7% of the total, and when leaving the market, 64.0% of the total is traceable. Yet, only 33.5% of the total is totally traceable in the whole process of before and after the wholesale stage, and only as little as 21.8% that is purchased by suppliers in the place of production is totally traceable. The main reasons are as following:

6.1.1 The Construction of Wholesale Markets Lacks Planning and the Market Development is Uneven

In China, the development of vegetable wholesale markets is accompanied by the development of urban and rural economies. Due to the lack of unified planning, and uneven development and unreasonable layout of wholesale markets, the funds from the government for the construction of wholesale markets are decentralized, which is one of the reasons why related government departments proclaim the lack of funds, and which, in addition, also leads to fierce market competition between different wholesale markets for consumer sources, as a result, some market management measures cannot be put into effect, and there is the lack of an effective management mechanism over the market. The survey also shows that for the current level of wholesale markets, the government can only directly control those state-owned or collective ownership wholesale markets, and lacks an effective management mechanism over private wholesale markets; therefore, the government cannot coordinate the behavior of all wholesale markets. On the other hand, suppliers do not fully understand the traceability system.

6.1.2 Low Level Market Competition Results in the Fact that the Traceability Management of Business Operators in Wholesale Markets is not in Place

Because all wholesale markets are self-governed, and in order to attract more customers and supply of goods and avoid the loss of customers due to the strict implementation of a traceability system, they will relax the implementation of relevant systems. First, there is no good implementation of the admission registration system. According to interviews with market managers, almost all wholesale markets have already established an admission registration system. However, according to the statistics of 410 questionnaires on suppliers, only 39.2% of all wholesale markets have implemented a vegetables admission registration system; Specifically, about 63.4% of municipal wholesale markets do so, but only 15.0% at the county level; In addition, only 53.5% of all wholesale markets implement a vegetables departure registration system, that is, about 81.2% of municipal wholesale markets are doing so, but only 25.8% are at the county level. Second, the proportion of vegetables market sampling is not high. The survey data indicates that only 7.1% of all suppliers receive a daily market sampling, 19.3% of all suppliers receive a market sampling 5–6 times a week, 24.5% of all suppliers receive a market sampling 3–5 times a week, and as high as 47.5% of all suppliers receive a market sampling 1–2 times a week. Third, there is no full implementation of the punishment mechanism. Although vegetable wholesale markets have penalties for unqualified vegetables, yet according to the survey, only 31.5% of all wholesale markets implement punishment mechanisms for unqualified vegetables.

6.1.3 The Source of Vegetables is Complex, and it is Difficult to Achieve Traceability Management

There are mainly three kinds of suppliers: suppliers who produce and market vegetables all by themselves, suppliers who purchase vegetables in places of production, and suppliers who are from upstream wholesale markets. The vegetable species and scale of such suppliers vary a lot, so the wholesale market operators need to adopt different quality and safety traceability systems. What's more, the suppliers of most business operators of wholesale markets are from other places, having higher mobility and a lower educational level. Therefore, the cost of the government implementing a quality and safety management system is high, but the effect is not good, which increases the difficulty for wholesale market traceability management. From the perspective of suppliers, the most difficult issue for vegetables traceability management rests with the big varieties of vegetables and the small scale of upstream vegetable producers, who are not good at recording and analyzing relevant information; the cost of traceability is high and it is difficult to implement the punishment mechanism on upstream producers.

6.1.4 Funds and Administrative Management System is not Favorable

The management of the supply chain of vegetables involves many departments, there are no unified supervision standards on vegetables entering wholesale markets, the functions of relevant supervision departments overlap, and the coordination mechanism is not perfect. For example, in farm product markets, relevant trade departments have not implemented strict systems for claiming certificates or invoices, resulting in the upstream suppliers or producers having no initiative to offer invoices.

6.1.5 The Effect of Supplier Training and Advocacy is not Good

In recent years, in order to strengthen the quality and safety management of agricultural products, relevant government departments have invested a lot of manpower and material resources to carry out supplier training and advocacy about the quality and safety management of agricultural products, but the training effort is not good and the training cost is high because of the high mobility of suppliers in wholesale markets. Therefore, the relevant training and advocacy aiming at vegetable suppliers are seriously inadequate, which directly results in suppliers' low level of awareness and recognition of the traceability system and its role. The survey results show that as high as 66.1% of all suppliers do not participate even one time in market training each year, representing 59.9% of municipal wholesale markets and 73.3% of county-level wholesale markets. Only 25.4% of the existing suppliers claim that they understand the traceability system; about 54.7% indicate that they only "have heard of the traceability system"; when asked "whether the traceability system can help to enhance the level of vegetable quality and safety", 47% of the suppliers express that they have "never thought about it". In the actual investigation, we further interviewed some suppliers, who have gained certain understanding of the traceability system, but many suppliers do not know in what way a traceability system is actually implemented, and they are not clear what measures are for wholesale markets and what is the purpose for implementing a vegetables quality and safety traceability system.

6.2 Analysis of Factors that Influence Suppliers of Vegetables Wholesale Markets Implementing a Traceability System

6.2.1 Selection of Measurement Methods and Variable Definitions

In recent years, the academic achievements in the study of quantitative economics on choice behavior and decision-making desire appear continually, and of which the application of logistic regression is most popular. Because the dependent variable, i.e. the suppliers, of this study only have two types of behavior, namely, "have implemented traceability system" and "have not implemented traceability system", this study adopts a binary logistic regression.

$$\ln\left(\frac{p}{1-p}\right) = \alpha + \sum_{k=1}^{n} \beta_k X_k$$

where, p is the probability of suppliers having implemented a traceability system, $\ln\left(\frac{p}{1-p}\right)$ is the logarithm of odds of suppliers having implemented a traceability system, α is the regression intercept, n is the number of independent variables, X_k refers to the kth independent variable, and β_k is the regression coefficient of the kth independent variable. The definiton of variables, the confidence interval, and the meaning of valuation are shown in Table 6.1.

Table 6.1 Definition of model variables

	Variable meaning	Meaning of valuation
Actual behavior	Whether using invoices or certificates when purchasing and selling vegetables or not	0=implemented, 1=not implemented
Ways of acquisition	Ways of acquire vegetables	1= produce on one's own, 2=from upstream wholesale market, 3= purchase in places of production
Quality certification	Whether the vegetables have gained certain certification or not	0=no, 1=yes
Types of customers	The main types of customers	1= households and mess halls, 2= supermarkets and farm product markets, 3= downstream suppliers
Customer attention	Whether most customers ask about the source of vegetables or not	0= no, 1= yes

(To be continued)

(Table 6.1)

Customers asking for invoices	Whether most customers ask for invoices and business vouchers or not	0＝no, 1＝yes
Admission registration	Whether wholesale markets have implemented vegetables admission registration system or not	0＝no, 1＝yes
Market inspection frequency	Times of sampling inspection conducted by wholesale markets each week	1＝0 time a week, 2＝1 to 2 times a week, 3＝3 to 4 times a week, 4＝5 to 6 times a week, 5＝7 times a week
Market penalty	Whether wholesale markets have come down upon unqualified vegetables or not	0＝no, 1＝yes
Departure registration	Whether wholesale markets have implemented vegetables departure registration system or not	0＝no, 1＝yes
Market training frequency	Times of training provided by wholesale markets for suppliers each year	1＝0 time a year, 2＝1 to 2 times a year, 3＝over 2 times a year
Government inspection frequency	Times of sampling inspection conducted by government departments each month	1＝0 time a month, 2＝1 time a month, 3＝2 to 3 times a month, 4＝4 and over 4 times a month
Government penalty	Whether government's penalties have influenced suppliers or not	0＝no, 1＝yes
Government training	Whether government departments have provided related training for suppliers or not	0＝no, 1＝yes
Age	The age of suppliers	1＝below 30, 2＝30 to 39, 3＝40 to 49, 4＝50 to 59, 5＝60 and above
Education	The educational level of suppliers	1＝primary school and below, 2＝junior middle school, 3＝senior middle school or technical secondary school, 4＝college for professional training and above

(To be continued)

(Table 6.1)

Years of service in selling vegetables	Years of suppliers selling vegetables	1 = 3 years and below, 2 = 4 to 6 years, 3 = 7 to 9 years, 4 = 10 to 12 years, 5 = 13 years and above
Income ratio	The ratio of the income from selling vegetables to the total household income	1 = 20% and below, 2 = 40% and below, 3 = 60% and below, 4 = 80% and below, 5 = 100% and below
Scale of operation	The scale of operation of suppliers compared in local wholesale markets	1 = low, 2 = relative low, 3 = medium, 4 = relative high, 5 = high
System awareness	The level of awareness of suppliers on vegetables quality and safety traceability system	1 = never heard of, 2 = heard of, 3 = general awareness, 4 = full awareness
Peer influence	Whether the decision of a supplier is influenced by other suppliers or not	0 = no, 1 = yes

6.2.2 Model Estimation of Vegetable Suppliers Implementing a Traceability System

In this study, we use SPSS16.0 statistical software to carry out binary logistic regression on survey data. The regression process involves the backward scalping method. In this process, the author respectively substitutes all independent variables, which influence the dependent variable, into the model, and then according to the test results, the author finds out the independent variable that has the most insignificant influence on the dependent variable. If the probability of the independent variable's coefficient being zero is significantly greater than 10%, this independent variable in the model should be removed. Then the author substitutes the remaining independent variables into the model to continue the test, and then repeats the aforementioned judgment, until the influence of all the independent variables on the dependent variable is significant.

The inspection process has gone through 14 steps. As we can see from the test results in Table 6.2 and Table 6.3, 13 independent variables have been successively removed, including Customers asking for invoices, Customer attention, Market training frequency, Education, Age, Government penalty, Departure registration, Market penalty, Peer influence, Types of customers, Government training, Quality certification, and Market inspection frequency. The statistic value of model goodness of fit $-2LL$ is 328.752, and Cox and Snell R^2 and Nagelkerke R^2 are 0.425 and 0.512, respectively, while the significance probability of the Hosmer-Lemeshow statistic is over 0.05, so the evidence is insufficient in this test to reject the null hypothesis. Overall, the goodness of fit falls into an acceptable range.

Table 6.2 Model estimation of vegetable suppliers implementing traceability system (initial)

	Regression coefficient	B index	Wald statistic	Probability of coefficients being zero significantly
Actual behavior	1.21***	3.385	19.133	0.000
Ways of acquisition	−1.184***	0.306	8.648	0.003
Quality certification	−0.605**	0.546	6.105	0.013
Types of customers	−0.057	0.945	0.027	0.869
Customer attention	−0.070	1.073	0.034	0.853
Customers asking for invoices	0.822**	2.274	4.576	0.032
Admission registration	−0.115	0.891	0.441	0.506
Market inspection frequency	−0.014	0.986	0.002	0.964
Market penalty	0.967**	2.631	4.453	0.035
Departure registration	0.060	1.062	0.037	0.847
Market training frequency	0.450**	1.568	6.512	0.011
Government inspection frequency	0.194	1.214	0.346	0.556
Government penalty	0.888*	2.431	3.632	0.057
Government training	−0.068	0.934	0.183	0.669
Age	0.062	1.064	0.088	0.767

(To be continued)

(Table 6.2)

Education	0.342	1.407	6.488	0.011
Years of service in selling vegetables	0.255	1.290	2.503	0.114
Income ratio	-0.053	0.948	0.117	0.732
Scale of operation	-0.466^{**}	0.627	5.950	0.015
System awareness	-0.528	0.590	1.535	0.215
Constant term	-3.675^{**}	0.025	5.722	0.017

Note: *, **, and *** indicates that the statistical tests are on 10%, 5% and 1% significance level.

Table 6.3 Model estimation of vegetable suppliers implementing traceability system (final)

	Regression coefficient	B index	Wald statistic	Probability of coefficients being zero significantly
Ways of acquisition	-0.597^{***}	0.551	12.691	0.000
Admission registration	1.024^{***}	2.783	14.255	0.000
Government inspection frequency	0.541^{***}	1.718	15.157	0.000
Years of service in selling vegetables	0.286^{***}	1.331	6.744	0.009
Income ratio	0.262^{*}	1.299	2.719	0.099
Scale of operation	0.244^{*}	1.277	2.929	0.087
System awareness	0.406^{**}	1.501	4.979	0.026
Constant term	-5.016^{***}	0.007	22.323	0.000

Note: *, **, and *** indicates that the statistical tests are on 10%, 5% and 1% significance level.

6.3 Results and Discussions

(1) Source of vegetables. "Ways of acquisition" has a reverse impact on the behavior of suppliers implementing a traceability system. The reasons rest with the cost of identifying vegetable information and that of implementing a vegetables quality and safety traceability system. Farmers, who produce and market their vegetables on their own, know very well the process for growing vegetables, and they are confident in the quality and safety of their vegetables sold in wholesale markets, so that they are in favor of a vegetables quality and safety traceability system. However, as for suppliers, who purchase vegetables in places of production, they have several and even dozens of individual providers, so it is difficult for them to know the quality and safety information of all vegetables; that is, the cost of identifying vegetable information and that of implementing a vegetables quality and safety traceability system is very high.

"Quality certification" does not have significant impact on the behavior of suppliers implementing a traceability system. This is because in recent years, our government relies on production bases and focuses on supporting leading agricultural enterprises and cooperatives in the implementation of relevant standards. Most of the vegetables entering the vegetable wholesale markets have received at least the certification of "Pollution-Free Agricultural Products".

(2) Whereabouts of vegetables. Through econometric analysis, we find that "Types of customers", "Customer attention", and "Customers asking for invoices" do not have significant impact on the behavior of suppliers implementing a traceability system. Although the main customer groups of various suppliers are different, each supplier will strive to maintain its credibility with fixed customers; therefore, different "types of customers" does not affect the behavior of suppliers implementing the traceability system. In addition, regardless of whether the majority of customers are asking about the quality information of vegetables or not, the suppliers will tell customers that the vegetables are safe and of high quality. In the case of the absence of fre-

quent market inspection, "Customer attention" does not have significant impact on the behavior of suppliers implementing a traceability system. Finally, in addition to supermarkets, farm product markets and other retail markets do not attach much importance to the management of business operators providing invoices for customers, and they even have no invoices, so that "Customers asking for invoices" does not have significant impact on the behavior of suppliers implementing a traceability system. In short, the purchasers in wholesale markets do not have an expected impact on the behavior of suppliers implementing a vegetables quality and safety traceability system, which indicates that the management of China's retail markets are not perfect and Chinese consumers' awareness of food quality and safety traceability should be continually enhanced.

(3) Market management. Because "Admission registration" records the producers, source, variety, and other useful information of vegetables, which clarifies the responsibility of potential punishment, vegetable suppliers may more actively implement the traceability system under the impetus of external factors. "Market inspection frequency" and "Market training frequency" do not have a significant impact on the behavior of suppliers implementing a traceability system. This is mainly because wholesale markets are "economies", and they attach more importance to their own economic interests. Therefore, in order to attract more customers, the market tends to reduce the number of inspections and relax the enforcement of punitive measures. That is why "Market penalty" and "Departure registration" have no significant impact on the behavior of suppliers implementing a traceability system.

(4) Government regulation. "Government inspection frequency" has significant positive impact on the behavior of suppliers implementing a traceability system, while "Government penalty" and "Government training" do not have significant impact. This is because, on one hand, government departments do not impose appropriate punishment on unqualified products, only very few of which are destroyed or confiscated, and the more common practice is that the suppliers are expelled out of the market. These suppliers will usually transfer this batch of unqualified vegetables to other wholesale markets. Therefore,

the government penalties do not have much impact on most suppliers. On the other hand, the training by government departments on suppliers has not been well enforced, because most suppliers are a floating population. So, it can be seen that due to low efficiency of management, government policies and measures do not play their expected role.

(5) Supplier characteristics. "Years of service in selling vegetables", "Income ratio", "Scale of operation", and "System awareness" have significant positive impact on the behavior of suppliers implementing a traceability system. With the lengthening of years of suppliers selling vegetables, they are more likely to accept the vegetables quality and safety traceability system. The higher the ratio of income from selling vegetables to the total household income is, the more likely the suppliers are to implement the traceability system, because the system can help them reduce the risks of substantial reduction in household income due to unsafe events. Similarly, the bigger the scale of the operation of suppliers in local wholesale markets, the more serious loss the suppliers will suffer from due to food safety issues; therefore, suppliers whose scale of operation is relatively bigger are more likely to implement the traceability system. In addition, the higher the awareness of suppliers on the traceability system, the more likely they are to rationally judge the long-term gains from implementing this system, and therefore they are more likely to utilize the traceability system. "Age", "Education", and "Peer influence" do not have significant impact on the behavior of suppliers implementing the traceability system. Suppliers of wholesale markets begin the vegetable business basically at different ages; some start their business in their teens, while some will start in their forties; "Age" is not a factor that can influence suppliers performing the traceability system. In addition, the author finds that the education level of most suppliers does not vary much, centering on the junior high school level. In actual practice, the awareness of suppliers of the vegetable industry and the various systems is more likely to affect their business behavior rather than their academic qualifications. Furthermore, the suppliers who are from different places in China are not fixed in one place and they do not frequently communicate with each other.

6.4 Policy Recommendations

6.4.1 Regulate the Construction of Agricultural Products Wholesale Markets and Improve the Circulation Pattern of Wholesale Markets

The government is duty-bound to ensure food quality and safety and consumer health and should make the appropriate expenditure for the construction of wholesale market traceability systems, but currently it may be difficult for the government to ensure the financial support for all wholesale markets. Therefore, the government should support and help to improve the market based on the degree of development of wholesale markets, such as preferentially and actively guiding public and collective ownership wholesale markets that are running well to first implement the traceability system and then gradually including private wholesale markets moving forward. As for investment projects, the government should give more priority to projects that not only are conducive to improving the means of implementing the quality and safety traceability system, but also can enhance the means for exchange and thus also innovate upon the different ways for exchange. In addition, the government can also utilize the land input, appropriately hold wholesale markets, and strengthen its supervision and management of wholesale markets, so that the wholesale markets can carry out the establishment of a agricultural products quality and safety traceability system according to government planning. After the infrastructure construction of agricultural products wholesale markets has been improved through these measures, the government should also guide wholesale markets to innovate upon and extend their service functions, encourage outstanding wholesale markets to carry out chain business, and encourage wholesale markets to expand from market functions to production, processing, packaging, storage and transportation, preservation, distribution and other related fields. Wholesale markets should also build related auction centers and electronic settlement centers. All these measures should help improve the business pattern of wholesale markets and lay a foundation for the effective operation of a traceability system.

6.4.2 Strengthen Collaboration between Government Departments and Establish a Unified Management Institution Consisting of a Couple of Departments

In order to overcome the drawbacks of multiple management and strengthen the communication and collaboration between various government departments, it is feasible to establish a management committee, composed of personnel selected by relevant functional departments of agriculture, trade, commerce and industry, and public security. On the basis of clearly defining the internal relations and job responsibilities, the management committee can coordinate and carry out real-time monitoring on the admission inspection of vegetables and on the implementation of requesting certificates and invoices, so as to ensure the quality and safety of agricultural products.

6.4.3 Strengthen Publicity and Training and Enhance the Enthusiasm of all Participants to Participate in the Traceability System

The government should step up the promotion and training of quality and safety traceability systems in all stages of production, circulation and consumption, enhance the awareness of all participants on the supply chain to quality and safety traceability, and guide them to initiatively coordinate and participate in the traceability system. It is particularly necessary to strengthen the role of media on the publication of the knowledge of quality and safety, on the cultivation of customers' awareness of quality and safety traceability, and on raising customers' awareness in requesting certificates and invoices and their willingness to pay for traceable agricultural products, so as to guarantee the operation of a traceability system through market-based instruments.

6.4.4 Actively Promote the Construction of Vegetable Production Bases and Cooperatives and Improve the Efficiency of Source Traceability

It is difficult for suppliers, in particular, those who directly acquire vegetables from the places of production, to master the quality information on the acquired vegetables, so the cost and risk of performing the traceability system is high. To this end, it is necessary to accelerate the construction of cooperatives and production bases for vegetables, and give full play to the unified management and service functions of cooperatives and production bases to improve the efforts of monitoring vegetable quality and safety and to enhance the integrity and normative records of production information. These measures can not only reduce the cost and risk of suppliers performing the vegetables quality and safety traceability system, but are also conducive to the innovation of the business entities in wholesale markets, to the cultivation of a combination of transportation and sale, and to the development and perfection of a vegetable market access system.

6.4.5 Accelerate the Process of Marketing Legislation and Promote the Legal Construction of Wholesale Markets

On the basis of domestic legislative research, investigating and learning from foreign laws and regulations, China should formulate as soon as possible a Law of the People's Republic of China on Agricultural Products Wholesale Markets and other relevant laws and regulations, so as to regulate the purpose of starting agricultural products wholesale markets, market planning, market access, trading rules, the duties of related employees, the structure and functions of administrative organs, and penalty provisions, etc. At present, since China has not formally promulgated such a law, local authorities can develop local approaches for the management of agricultural products wholesale markets. Local governments at all levels should speed up the legislative work for the management of local wholesale markets planning and construction, implement macro-control over construction of wholesale markets, develop medium and long-term planning for the development of wholesale markets, and supervise investors to construct modern and high-level wholesale markets.

References

Banterle, A. & Stranieri, S. (2008). The consequences of voluntary traceability system for supply chain relationships. An application of transaction cost economics. *Food Policy*, 33(6), 560-569.

Caswell, J.A. (1998). Valuing the benefits and costs of improved food safety and nutrition. *Australian Journal of Agricultural and Resource Economics*, 42(4), 409-424.

De Castro, P. (2002). Mechanisation and traceability of agricultural products: a challenge for the future. *Journal of Scientific Research and Development*, 8, 120-129.

Fan, F.R. & Jin, H. (2006). Strategic study of origin mechanism on establishment of agricultural product quality security in Wuhan. *Journal of Huazhong Agricultural University (Social Science Edition)*, 93(4), 50-52.

Golan, E.H. (2004). Traceability in the US food supply: economic theory and industry studies. US Dept. of Agriculture, Economic Research Service.

Golan, E.H., Vogel, S.J., Frenzen, P.D. & Ralston, K.L. (2000). Tracing the costs and benefits of improvements in food safety: the case of hazard analysis and critical control point program for meat and poultry. *Agricultural Economics Reports*.

Golan, E.H., Krissoff, B., Kuchler, F., Nelson, K.E., Price, G.K. & Calvin, L. (2003). Traceability for food safety and quality assurance: mandatory systems miss the mark. *CAFRI: Current Agriculture, Food and Resource Issues*, (04), 27-35.

Golan, E.H., et al. (2004). Traceability in the US food supply: economic theory and industry studies. US Dept. of Agriculture, Economic Research Service.

Golan, E.H., Krissoff, B. & Kuchler, F. (2005). Food traceability: One ingredient in a safe and efficient food supply. *Prepared Foods*, 2(2), 59-70.

Hall, D. (2010). Food with a visible face: Traceability and the public promotion of private governance in the Japanese food system. *Geoforum*, 41 (5), 826-835.

Hobbs, J.E. (2004). Information Asymmetry and the Role of Traceability Systems. *Agribusiness: An International Journal*, 20, 397-415.

Hobbs, J.E., Agriculture, C. & Canada, A.F. (2007). Identification and Analysis of the Current and Potential Benefits of a National Livestock Traceability System in Canada. Agriculture and Agri-Food Canada.

Chapter

7

Investment in Voluntary Traceability: Analysis of Chinese Hog Slaughterhouses and Processors

7.1 Introduction

The production and consumption of agricultural products has gradually come to comprise a complete "farm-to-table" traceability management system. Due to the credence attributes of agricultural products that consumers cannot know even after consumption, the information on food safety and nutrition is of importance for consumers from the upper suppliers, including hormones, antibiotics, salmonella infections, chemical pesticide residues of food safety, nutrition content and proportion relevant to nutrition and health, origin of country, animal welfare and the like (Golan *et al.*, 2005). Traceability is becoming an internationally acceptable method of providing safer food supplies and of connecting producers and consumers. On March 15th, 2011, Jiyuan Shuanghui Food Co., Ltd. reported that it purchased pig feed containing Clenbuterol, a chemical which can be fed to pigs to prevent them from accumulating fat. It is banned as an additive in pig feed in China for it can end up in the flesh of pigs and is poisonous to humans if ingested. This firm is famous for its quality control system — "eighteen inspection procedures ensuring quality of its pork meat". However, the additive scandal would severely damage its brand image, also impacting the whole industry. Consumers began to doubt about the safety of pork meat they bought whether from a supermarket, wet market, or collective shop. In just one week after the case was reported, the economic loss amounted to ¥30,000,000 (nearly USD $4,639,230) from reduced pork sales across the entire live pig raising industry (Changchun Evening News, 2011).

China is the world's largest pork producer and consumer, accounting for 50.2% and 50.4% of the total world's pork meat in 2010, respectively. However, with a gradual increase in pork safety incidents, the public is experiencing wider concerns about pork quality. The problem related to pork safety and quality is in essence related to imperfect information, leading to market distortion and morality. Market failure in the food market is commonplace (Caswell, 1998; Golan *et al.*, 2003; Ritson and Mai, 1998). However, the information on food safety not renewed in time mostly results from lack of government's disclosure transparency (Blue Book of Rule of Law, 2011).

Therefore, many experts advocate that the mechanism of "quality ensuring good price" and a punishment system based on clear responsibility as well, to some extent would motivate hog slaughterhouses and processors (hereafter, HSPs) or retailers to provide enough information referring to food safety for consumers, thus reducing the market failure. In such circumstances, mandatory traceability has been introduced as an efficient tool to facilitate the imperfect information, and to ensure the removal of unsafe food from the food supply chain. For example, the UN required that all meat be tracked and traced, or else it would be banned from the market, and this law has been in place since January 1st, 2005. In Japan, the good agricultural product's certification system has been introduced to identify every product in Japanese markets during the same year. The Chinese government has established the systematic identification or traced back to origins of agricultural food in some developed areas, so as to improve food standards related to production and distribution, and to provide more information for consumers since 2002. But not until 2010, were the first 20 cities[1] selected for the pilot project (promulgated by the government but not required) to adopt distribution traceability in the meat and vegetable industry by China's commerce ministry. The adoption of traceability will no doubt add cost to the information distribution among producers, distributors and processors, according to the *Basic Requirement of Traceability System on Meat Distribution* issued by Ministry of Commerce in 2011. Some chose not to renew their traceability system, while others applied for renewal traceability that has more stringent standards and controls along the supply chain based on market or other factors. This raises questions about the underlying drivers of investment in voluntary traceability and implications of access to large wholesale markets or super wet markets, where traceability is required.

Technologies supporting the system for traceability are becoming available now, while the profitable feasibility of such systems is the main point. As a result, traceability characteristics such as cost of traceable behavior, and the equilibrium distribution structure should be taken into consideration. Under the Hypothesis of Economic Man, the manager from HSP must first balance its benefits and costs when adopting safety traceability. Such cost-benefit balance from traceability has always been one of the hottest topics in theory and in practice, especially about hidden costs and benefit.

We try to study whether the motives with the introduction of such a system would bring discrepancy between the benefits and transaction costs, and would lead to an available approach to encourage a system-wide adoption of voluntary traceability. The distribution of questionnaires was also done to obtain first-hand data and comparative analysis was also performed to confirm our findings. Our results indicate that transaction cost savings integrated up/

---

[1]Twenty cities are Shanghai, Chongqing, Dalian, Qingdao, Ningbo, Nanjing, Hangzhou, Chengdu, Kunming, Wuxi, Tianjin, Shijiazhuang, Harbin, Hefei, Nanchang, Jinan, Hankou, Lanzhou, Yinchuan and Urumqi City in China and are supported by the central finance department.

downstream, customer specialized requirements, and consumer awareness of food safety are most significant motivations influencing investment in voluntary traceability by HSPs in China.

The rest of this chapter is organized as follows. Section 7.2 provides a brief review of related economic literature. Section 7.3 describes the data and materials used in our analysis. Section 7.4 presents the results and discussion. Section 7.5 presents the conclusions.

7.2 Related Literature

A wide range of relevant research on agricultural product traceability is based on transaction cost theory and cost-benefit analysis. Hobbs (1996) studies show that adoption of pork traceability increases the transaction cost of beef slaughterhouses when choosing their supplier channel. The cost-benefit analysis is used to determine that traceability management efficiency depends on a private firm's costs-benefits balance (Golan *et al.*, 2004). Other factors such as width, depth and precision of traceability, complexity of the system and technical system may affect the traceability cost, which also differs among various products (i.e. fruit and vegetables, aquatic products and meat products), therefore, researches assume specific product or industry in their studies.

The introduction of product traceability offers the basis for information transfer among supply chain partners. Its tracing and tracking capability ensures production and supply of safe products, reduction in food-safety incidents, and therefore strengthens public health. These attributes bring positive external results, causing the government to likely support costly and burdensome traceable behavior of HSPs or farmers (Golan *et al.*, 2003; Hobbs *et al.*, 2002).

Agricultural product traceability is high asset specificity, making its unique resource an advantage by specific investment. Nevertheless, this investment in traceability is relatively large. Narrod *et al.* (2009) identifies the scale of economy in conducting a traceability system. It is an effective way to tackle this high-cost problem by a combination of government, flagship enterprises and farmers collective activities, which a single agricultural farmer could hardly afford. In order to respond to market failure, the government regulation aims at social welfare maximization from food safety/quality and a private firm's profit maximization. From this point of view, the extent of prefectural and local government support in helping to incentize HSPs to conduct traceability is one of the key issues in this study. Introduction of government-monitored traceability can facilitate governmental surveillance and domestic agricultural product trade. However, mandatory traceability for all agricultural food may not be the most efficient mechanism for verifying quality claims (Golan *et al.*, 2003).

In existing studies, many researchers (Fan and Jin, 2006; Lei, 2010; Li *et al.*, 2004; Ye *et al.*, 2011; Zhou and Jiang, 2007) propose to strengthen governmental support for industrialized agricultural organizations (Zhou and Jiang, 2007), in view of the current problems from adoption traceability, by increasing the incentive to adopt traceability, based on the present food safety and quality management and technology system. Such systems like GAP, GMP and HACCP can contribute to establishing a fast trace back monitoring system of information documenting for the consumer, and the government concerned. However, the references have no further discussions on how to support and extend this system.

Government-mandated traceability systems (Hall, 2010) prescribe one traceability template that requires all HSPs to adopt, or they will be fined. It is in some cases necessary to ensure food safety and market supervision, to protect a consumer from fraud and HSPs from imperfect competition with low-quality product's HSPs. As Souza-Monteiro and Caswell (2004) stated, the adoption of traceability would ensure animal and human health as well as identify the firm's responsibility throughout the supply chain when food safety issues occurred. Some detailed information (e.g., incoming/ante-slaughter inspection, synchronized quarantine, and outgoing record) by safety standards along the supply chain's key stages such as production, wholesale, distribution and retailing, will be provided to consumers. However, every raw material, input, and process makes an enormous and costly task (Golan *et al.*, 2003). Few, if any, firms or consumers would be interested in all such information; thus, some of them were excluded.

A voluntary traceability system, developed by private sectors and not formally required, is introduced regarding a firm's brand image, credence attributes of products and long-run benefits in order to increase higher product quality and to substantiate their credibility claims through traceability. Considering it as an institution, voluntary traceability is a more flexible choice. In conclusion, we are interested whether there exists a profound discrepancy in investment between the above two traceability systems, and its underlying drivers.

7.3 Materials and Methods

This chapter focuses on the adoption of a Chinese traceability system and the impact of adoption on prices, input and output quantities and costs based on a survey of HSPs conducted from July to September in 2010 across Zhejiang Province, located in the eastern region of China. Names and addresses of potential respondents were obtained from the Meat Association. Among the list of 120 pork slaughterhouses and processers[2], twenty-two HSPs' managers agreed to participate in a personal interview and agreed that the authors can visit their HSPs/plants. Ninety-eight other questionnaires were sent by e-mail after we phoned the firm managers who agreed to participate in the survey. The following groups were selected to cover the main activities related to the traceability system in HSPs: supply chain managers, chief executive officers and decision makers.

---

[2] Some branches of slaughterhouses and producers belong to the consolidated slaughtering plants and processing firms, and as such, they adopt the same traceability system (i.e., mandatory or voluntary traceable behavior) as their parent firms do.
"For food trading enterprises that adopt a centralized distribution model, the headquarters of the enterprises may centrally check the license of the supplier and compliance certificates of the food and create an inspection record for incoming food products" (Article 39 of the Food Safety Law of the People's Republic of China, 2009).

In our approach, the questionnaire consisted of four parts. Part 1 contained questions about the general features of the HSPs, i.e. the numbers of employees, educational background of managers, a firm's legal status, and business type. Part 2 consisted of several questions about the cost categories of investment in traceability which were drawn primarily from Maldonado et al. (2005). Also included was the private benefit from traceability, in particular, what the government can do to motivate a firm's further investment in voluntary pork traceability. Part 3 focused on the manager's motivations to adopt traceability. Some questions were related to the variation in a firm's asset specificity of transactions related to the adoption of voluntary traceability. We used some proxies to access differences among twelve motives. A five-point Likert scale was used ranging from not important (1) to extremely important (5) in order to report motivations of a firm's manager to invest in the voluntary traceability system. A high score suggested a high probability of adopting a voluntary traceability system.

We defined voluntary traceability in combination with the definition of traceability, i.e., breadth, depth, and precision (Golan *et al.*, 2004). Breadth represents the amount of information collected from the quality attributes of pork and its byproducts. We specify that the firms providing quarantine standards beyond the standards from a mandatory system have voluntary traceability in China. Depth of traceable behavior is defined as how far back or forward the behavior tracks the relevant information in a food supply chain. Meanwhile, precision refers to the degree of assurance, with which the tracing behavior can identify the food safety attributes of pork or its products. When the objective of a traceability system is extensive, take the whole farm as an example, the precision regarding safety and quality is a little inefficient, but when referring to a single cow or pig, it can pinpoint a particular objective quality or characteristic. In our study, heavy metal testing, a ratio of eight percent or above of the sample testing on the Clenbuterol additive or other animal medicine residues, are considered as voluntary traceable behaviors.

7.4 Results and Discussions

Key findings of the current study are divided into five sections. The first section addresses general evidence of the survey. The next section discusses the benefit and cost of traceability. The third section delineates the expected benefits from investment in traceability. The following section describes governmental policy and its relevant support related to the investment in the traceability system. The final section presents the motives for venturing into a pork traceability system.

7.4.1 General Evidence of the Survey

In total, our sample consisted of 120 HSPs. Ninety questionnaires were received by the end of October. Among these, nine were invalid as they contained quite incomplete information. A total of eight-one questionnaires were considered valid. Twenty-nine of these valid samples were HSPs which had introduced voluntary traceability, and the remaining fifty-two were those adopting mandatory traceability in practice.

With regard to the HSP scale, the sample consisted of both big and small HSPs, 53.8% of HSPs with employees less than 50, 20.5% from 50 to 100, 19.2% between 101 and 500, and the rest, 6.4%, with more than 501 employees (Table 7.1). And a large ratio of HSPs with voluntary traceability (13.8%) had more than 501 employees compared with that of HSPs who had implemented mandatory traceability (2.0%). Four-fifths of the HSPs with more than 501 employees conducted voluntary traceability, while a low ratio—nearly 33.3% of HSPs with less than 50 employees also did so. It was found that 58.5% of the sample were private HSPs, 32.5% were state-owned HSPs, and the remainder joint-venture HSPs. About 74.1% of HSPs adopting voluntary traceability were private HSPs, a higher percentage than those HSPs adopting mandatory traceability (50%). The number of county-level and municipal leading HSPs adopting mandatory traceability amounted to 36 (87.8% of the group). Half of HSPs with a national flagship and provincial flagship conducted voluntary traceability.

Table 7.1 Hog slaughterhouses and processors' features

	Mandatory		Voluntary		Total	
	Number	Share	Number	Share	Number	Share
	(*n*)	(%)	(*n*)	(%)	(*n*)	(%)
Employees						
<50	28	57.1	14	48.3	42	53.8
50-100	11	22.4	5	17.2	16	20.5
101-500	9	18.4	6	20.7	15	19.2
>501	1	2.0	4	13.8	5	6.4
Legal status						
State-owned	20	40.0	5	18.5	25	32.5
Private	25	50.0	20	74.1	45	58.4
Joint-ventured	5	10.0	2	7.4	7	9.1
Education background of manager						
Junior secondary school level	6	12.8	2	4.0	8	11.0
High school graduate	23	48.9	11	42.3	34	46.6
Undergraduate or above	18	38.3	13	50.0	31	42.5
Flagship grade						
State-level	1	2.4	1	4.5	2	3.2
Provincial level	4	9.8	4	18.2	8	12.7
Municipal level	16	39.0	5	22.7	21	33.3
County-level	20	48.8	12	54.5	32	50.8
Business type						
Slaughtering	46	52.3	17	42.5	63	80.8
Carcass meat	30	44.8	14	35.0	44	56.4
Cold fresh meat	4	4.9	3	7.5	7	8.9
Pork product	7	8.0	6	15	13	16.7
Brand grade						
State-level	1	2.4	2	8.0	3	4.5
Provincial level	5	12.2	7	28.0	12	18.2
Municipal level	6	22.0	7	28.0	16	24.2
No brand	26	63.4	9	36.0	35	53.0

Regarding educational background of managers, thirteen (50% of answers) from HSPs adopting voluntary traceability and eight (38.3%) from HSPs conducting mandatory traceability were undergraduate or above. The higher the education managers received, the more likely their HSPs were to introduce a voluntary traceability system.

Concerning the business type between HSPs adopting the mandatory and voluntary traceability, the results revealed that 80.8% of the HSPs operated pig slaughtering, 56.4% with carcass meat, 16.7% with pork processing and the remainder with cold fresh meat. Although most of the samples focused on both live pig slaughtering and carcass meat, three HSPs (7.5%) and six HSPs (15%) had cold fresh meat and pork products, while for these HSPs, government-mandated traceability was in place only 4.9% and 8%, respectively. However, for HSPs making cold fresh meat and pork products themselves, nine HSPs (45%) were conducting voluntary traceability. Sixteen HSPs, accounting for 74% of respondents' implemented voluntary traceability owned brands at a national, provincial and city level. Among these, the overwhelming majority of HSPs were provincial and city brand owners.

7.4.2 Benefit and Cost of Traceability

Traceability benefits and costs vary across HSPs, as well as the effect of breadth, depth, and precision of each traceability system. On the cost side, the firm's traceability costs consists of equipment and technical costs, labor costs for food-safety assurance, testing and traceability management (Golan, 2004; Lei, 2010; Meuwissen *et al.*, 2003; Ye *et al.*, 2011). Other researchers (De Castro, 2002; Golan *et al.*, 2000; Hobbs *et al.*, 2007; Souza-Monteiro and Caswell, 2004) point out that such factors as consumer preference, degree of traceable information, law's constraints, coordination approach between supply chain partners, traceability, technical level and an enterprise's strategy determine how enterprises try to balance benefits with costs of traceability.

This section of the survey regarding the calculation of costs underlines an introduction of traceability at the level of human, physical, geographical and intangible asset specificity. In this study, the total and the average cost of training charges, consulting fees, detection, information recording and documenting, etc., marketing management, and indirect management for application of traceability and information systems reorganized were listed in Table 7.2.

The survey underlined that the main costs of introducing the traceability in pork supply chains were equipment costs (RMB ¥ 1388.3×10^4 in total), followed by newly enrolled technical and management staff (RMB ¥ 543.5×10^4 in total) and detection costs (RMB ¥ 497.5×10^4 in total). To further discuss the differences between HSPs with voluntary traceability and mandatory traceability, the average cost of each category from the two groups was also calculated. The results showed that the average cost of equipment needed for voluntary traceability was far less than that of the mandatory traceability. An advantage of the application of voluntary traceability was that most of the infrastructure required to implement it already existed.

Table 7.2   Average total cost of adopting traceability system by slaughterhouses and processors (Unit: ten thousands RMB annually)

Cost category	Total cost		Mean cost	
	Mandatory	Voluntary	Mandatory	Voluntary
New staff	396.8	146.7	12.4	9.17
Training cost	75.7	73.9	3.29	6.16
Equipment cost	1298.7	89.6	44.8	6.4
Lab test cost	429.2	68.3	19.5	5.69
Consultant's cost	48.5	16.6	4.85	4.1
Information records & ducuments	133.4	47	5.56	5.88
Marketing	103.8	345	8	43.1
Indirect cost (Staff time in traceability design, and total operation costs)	199	90	7.65	8.18

Training charges included training suppliers, customers and technical staff. The average cost of training technical staff, customers and suppliers was nearly twice the cost of building a mandated system. A possible explanation may be that a higher-level motivation of integrating up/down agents along the supply chain on adopting voluntary traceability results in expanding the cost of training. A significant difference was the average marketing cost of implementing voluntary traceability, RMB ¥ 43.1×10^4 per annum, more than five times as much as that of those HSPs with mandatory traceability. With an increasing attention on food safety and quality from consumers, firms have adjusted their investment to focus more on marketing and paid more attention to its product credence attributes, to protect themselves from other fraudulant products or low-quality products.

7.4.3 Expected Benefits from Investment in Traceability

With regard to a firm's decision concerning expected benefits, the impact of private benefits comes to our mind. These benefits, some considered to be hidden, comprise the reduction in potential for safety issue losses, product-sales growth, operational management efficiency, brand improvement and potential reduction in information-coordinated costs along the pork supply chain.

Less evident were expected benefits connected to reduction in information coordination cost, only 26.3% of HSPs adopting mandatory traceability and 28.7% of HSPs adopting voluntary traceability stated that traceability had improved coordination efficiency with customers along the supply chain (Fig. 7.1). This could be explained in terms of stringent production and processing controls having prevented information flow to partners throughout a pork supply chain in practice. The results also showed that:

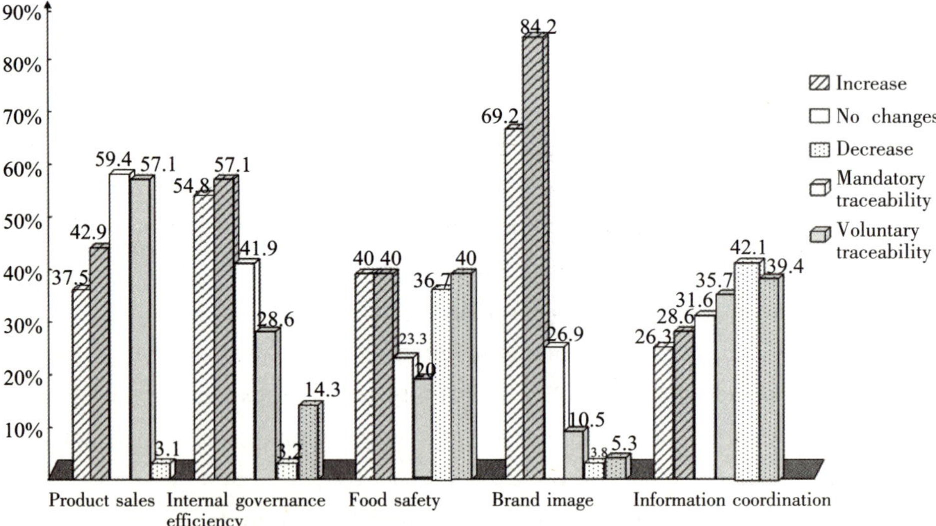

Fig. 7.1 Private benefits from adopting traceability system

• voluntary traceability was an important factor in a firm's brand improvement (84.2% which is more than 69.2% of HSPs with mandatory traceability);

• more HSPs (42.9% of the respondents) had annual sales growth from adopting traceability than that of HSPs (37.5% of respondents) who had introduced mandatory traceability.

• more HSPs conducting voluntary traceability had an efficiency drop for internal management (14.3%) than that of firms conducting mandatory traceability (3.2%).

A brand improvement could be expected to be due to a firm's investment in marketing. One possible explanation for this could be that investment in marketing, for HSPs adopting voluntary traceability, provides several channels, such as advertising, promotion and product presentation to verify their product with high-quality or other excellent quality characteristics to consumers, as consumers have awareness of food safety. However, as the stringent production rules and controls on the system are implemented, and the more frequent exchange of traceable information along supply chain, efficiency of management would be decreased to some extent.

7.4.4 Governmental Policies and its Relevant Support Related to the Investment in the Traceability System

In some circumstances, the amount of traceability systems implemented by the HSPs may be less than the social optimum, as the social and public benefits far exceed those of the firm's benefits from a traceability system. On the other hand, HSPs may not be willing to be exposed to liability and have an incentive to under-invest in traceability in order to maintain some level of anonymity. In order to achieve some kind of equilibrium, the traceable behavior should be stimulated by government financial support (Jin *et al.*, 2008). Firms with a voluntary traceability system, to some extent, expand the consumer's choices and protect them from fraud and unsafe food (Golan *et al.*, 2003).

Concerning the question "Which three important governmental initiatives do you expect to support investment on traceability?—technological training and guidance, fiscal subsidies, marketing promotion and knowledge of food safety for consumers, market information service and market standardization", the results showed that all respondents equally agreed on the fiscal subsidies, then market standardization, and consumer knowledge on food safety was lagging behind. In addition, there were some differences between HSPs with the introduction of voluntary traceability and mandatory traceability (Table 7.3).

A relative higher ratio of HSPs adopting a voluntary traceability system hoped to get more governmental support compared to those employing mandatory traceability. The first three highest ranked projects were financial support (92%), unifying standards of traceable product's market (76%) and consumer education on food safety and quality (52%). Government financial support might help a firm to facilitate applications of a more stringent traceability system management, which is very easy to understand. A possible explanation regarding market standardization was that unified standards could differentiate voluntary traceable products from fraudulent products or other lower-quality products.

Table 7.3 Governmental support expected by HSPs

Governmental support	Mandatory		Voluntary	
	Number (n)	Share (%)	Number (n)	Share (%)
Technological training	17	40.5	12	48
Financial support	32	76.2	23	92
Consumer education	18	42.9	13	52
Information service	5	11.9	4	16
Market standards	27	64.3	19	76

7.4.5 Motives for Venturing in the Pork Traceability System

The survey underlines the major motives for introducing a traceability system in pork slaughter and processing HSPs fast trace back capacity (the average score 4.49), incentives (promotions by) from associations and prefecture and local governments (4.13) and operating efficiency improvements (4.07) (Table 7.4). Asymmetric information problems in an agricultural food market may result in market failure (Golan *et al.*, 2003). A traceability system, whether mandatory or voluntary, allows HSPs to protect themselves from unfair competition and to pinpoint their production problems, thus minimizing the extent of recalls. In this scenario, HSPs choose a fast trace back capacity as the most important motive.

Table 7.4 Motives for investment in the traceability system.

Motives	Mean score		Total score	F-test
	Mandatory	Voluntary		
Operation efficiency improvement	3.94	4.29	4.07	2.502
Brand image improvement	3.83	4.29	4.0	2.552
Fast traceable capacity	4.40	4.64	4.49	1.478
Product differentiation	3.82	4.18	3.95	2.453
Transaction cost saving integrated up/downstream	3.70	4.15	3.86	3.180*
Incentive from association & government	4.02	4.32	4.13	1.613
Direct government supervision	3.45	3.74	3.55	1.125
Customer requirements on traceability	3.00	3.85	3.32	9.165**
Cost of application of traceability	3.00	3.33	3.12	1.524
Consumer awareness of food safety	3.38	3.89	3.56	3.390*
Competitor's safety and quality level	3.48	3.61	3.53	0.179
Requirement of laws and regulations	3.86	4.18	3.97	2.093

Note: *,**,*** represented significant at 10%, 5%, and 1% level, respectively.

A variance analysis (ANOVA) was then performed to verify whether there were any variations in the drivers to adopt voluntary or mandatory traceability. The results revealed that differences existed between the two groups on whether adopting voluntary traceability or not would lower a transaction cost of integrated up/downstream, complied with customer demand on traceable products and matched consumer demand on high-quality and safe food (consumer awareness of food safety). It is inevitable that HSPs who invested in voluntary traceability were more inclined to integrate with other agents in the supply chain for long-term cooperation and transaction cost saving. One possible answer is that HSPs could build credibility of their own quality product through traceability to identify their products during the supply chain.

This can be explained in terms of more stringent production and processing rules and controls, not in terms of product differentiation. HSPs introduced voluntary traceability for high-quality and safety assurance to target consumers. Aiming at a decrease in the degree of transaction uncertainty regarding the higher level of transaction transparency along the pork supply chain (Banterle and Stranieri, 2008), HSPs also needed to match demand of customers or upper/downstream suppliers who were closer to the end consumers, and then to expand specific investment for conducting voluntary traceability. A strict enforcement of responsibility might lead to an overwhelming supply of safety initiatives for inadequate quality products. No HSPs can provide a product with absolute safety. Kolstad *et al.* (1990) indicate that responsibility and laws and regulations are complementary. However, requirements of laws and regulations are not significant, as shown in our results. A possible reason is that, as Innes (1994) indicates, appropriate fines for responsibility might be more effective than those standards established in a food market with asymmetrical information.

7.5 Conclusions

The traceability system is becoming increasingly important for controlling and monitoring agricultural product safety. In the pork industry, introduction of a traceability system is at its first stage in China, as only some leading HSPs meet the requirements for a system, not to mention the need for voluntary traceability with higher standards in food safety and quality. For capital-intensive systems for application of traceability management and its asset specificity, a firm's benefits cannot outweigh the costs in the short run due to their small market share. In other words, market share will be improved, as well as a firm's benefits and cost balance, if the implementation degree of a food traceability system has been widely improved. From the comparative analysis and discussion in previous sections, a profile of pork slaughter and processing HSPs most inclined to invest in voluntary traceability is:

Medium and large-sized private HSPs with more than 100 employees. As Golan *et al.* (2003) stated, mandatory traceability may be burdensome and costly to adopt for all HSPs and all products, thus a flexible option of voluntary traceability is a good choice for a firm to undertake for credence character assurance. From this point of view, regulation should be set up for unifying standards in the traceable product market to attract private investment in voluntary traceability. At present, small-sized HSPs with less than 50 employees account for the majority of firms considering private investment in adopting voluntary traceability. Because of lack of capital, these HSPs are not able to invest in equipment, professional skills training and other resources to conduct voluntary traceability. Government should also expand private traceability choices for these HSPs, by introducing more feasible standards and by offering traceable support service and training.

Decision maker with undergraduate or higer education background. Some accessible approaches should be provided to higher decision maker's education, for instance, training in relevant knowledge of the pork industry or other kinds of continuing education.

Firm's operation in slaughtering and carcass meat production. The degree of industrialization in China's pork industry is relatively low, and advanced conditions for present pork industry concentration is also needed. The shortage of technical staff has existed in the processing business with strict product quality requirements, especially with the introduction of voluntary traceability. For development of pork industrialization and pork added value, nonprofit training for supplying the staff is insufficient and should be an organization initiative by local government, to encourage a flagship firm to expand its production scale. An agricultural industrialization mode of "pork HSPs-led+ scattered farmer" should also be promoted. A close link between HSPs with scatter farm breeders, i.e. setting up own hog breeding bases, is used to reduce the cost of raw material quality control and degree of information asymmetry to some extent, which is consistent with the ultimate purpose of traceability from "the farm to fork".

HSPs owned brands at a national, provincial and city level. HSPs, introducing voluntary traceability, were those brand's owners, which could correlate the survey results with the marketing cost. Branding in the pork industry produces higher costs, but it provides more gains than that. On one side, it brings private benefits for a firm from its image improvement and really creates brand equity, thus lowering the cost of marketing sooner or later. On the other side, it indeed helps consumers to decide on what higher-quality products to buy. When HSPs aim at long-lasting brand building, the brand itself provides a platform for information communication with the public to approach relative information equilibrium, and finally increases social benefits. For the government, in an effort to alleviate confusion in the market in pork competition, support comes into mind for HSPs to build a city or increase their brand influence.

Although motives such as compliance with customer demand on special standards and cost savings from integration with the up/downstream showed the significant difference between voluntary and mandatory traceability, HSPs cannot improve in coordination efficiency, to some extent, due to imperfect information throughout the pork supply chain. Strengthening government support for building a system wide traceable information flow, which can be freely shared by all HSPs and supply chain partners, would be a big benefit. Meanwhile, policy makers should carry out some measures to incentivize hog slaughter and processors involved in a traceability system extension, to guarantee pork safety and quality nationwide and, in addition finally improve social benefits.

Knowledge of food safety and quality voluntary traceability popularization is necessary by the government. The higher consumer awareness is on food-safety traceability, the more high-quality food is in demand. Such an expanded demand for food traceability can stimulate HSPs to invest in voluntary traceability to expand production. Moreover, financial subsidies for an HSP can facilitate the application of burdensome and costly traceability. Take marketing charges, for example; it was the main cost for these HSPs in implementing voluntary traceability. For a firm, this can be a supply for capital insufficiency to provide food safety and quality information to consumers, as a tool to control its traceability management and as its social responsibility (Zhou and Ye, 2007).

Above all, investment in adaption of voluntary traceability by HSPs in China will add a new practical guideline to the international pork industry. Also, the experience from a "pork slaughter and processor-led + farmers" mode in the concentration process of the pork industry and, in agricultural extension of pork safety and quality traceability, will be considered as a reference for other developing countries.

References

Banterle, A. & Stranieri, S. (2008). The consequences of voluntary traceability system for supply chain relationships. An application of transaction cost economics. *Food Policy*, 33(6), 560-569.

Blue Book of Rule of Law (2011). Institution of law of Chinese of Academy of Social Science. Beijing: Social Science Academic Press.

Caswell, J.A. (1998). Valuing the benefits and costs of improved food safety and nutrition. *Australian Journal of Agricultural and Resource Economics*, 42(4), 409-424.

Changchun Evening News, 2011. Http://ccwb.1news.cc/html/2011-03/22/content_137892.htm.

De Castro, P. (2002). Mechanisation and traceability of agricultural products: a challenge for the future. *Journal of Scientific Research and Development*, 8, 120-129.

Fan, F.R. & Jin, H. (2006). Strategic study of origin mechanism on establishment of agricultural product quality security in Wuhan. *Journal of Huazhong Agricultural University (Social Science Edition)*, 93(4), 50-52.

Golan, E.H. (2004). Traceability in the US food supply: economic theory and industry studies. US Dept. of Agriculture, Economic Research Service.

Golan, E.H., Vogel, S.J., Frenzen, P.D. & Ralston, K.L. (2000). Tracing the costs and benefits of improvements in food safety: the case of hazard analysis and critical control point program for meat and poultry. *Agricultural Economics Reports*.

Golan, E.H., Krissoff, B., Kuchler, F., Nelson, K.E., Price, G.K. & Calvin, L. (2003). Traceability for food safety and quality assurance: mandatory systems miss the mark. *CAFRI: Current Agriculture, Food and Resource Issues*, (04), 27-35.

Golan, E.H., et al. (2004). Traceability in the US food supply: economic theory and industry studies. US Dept. of Agriculture, Economic Research Service.

Golan, E.H., Krissoff, B. & Kuchler, F. (2005). Food traceability: One ingredient in a safe and efficient food supply. *Prepared Foods*, 2(2), 59-70.

Hall, D. (2010). Food with a visible face: Traceability and the public promotion of private governance in the Japanese food system. *Geoforum*, 41(5), 826-835.

Hobbs, J.E. (1996). A transaction cost analysis of quality, traceability and animal welfare issues in UK beef retailing. *British Food Journal*, 98(6), 16-26.

Hobbs, J.E., Fearne, A. & Spriggs, J. (2002). Incentive structures for food safety and auality assurance: an international comparison. *Food Control*, 13, 77-81.

Hobbs, J.E., Agriculture, C. & Canada, A.F. (2007). Identification and analysis of the current and potential benefits of a national livestock traceability system in Canada. Agriculture and Agri-Food Canada.

Innes, R. (1994). Liability rules and safety regulation under asymmetric information. Unpublished manuscript (Department of Agricultural and Resource Economics, University of Arizona, Tucson, AZ).

Jin, S., Zhou, J. & Ye, J. (2008). Adoption of HACCP system in the Chinese food industry: A comparative analysis. *Food Control*, 19(8), 823-828.

Kolstad, C.D., Ulen, T.S. & Johnson, G.V. (1990). Ex post liability for harm vs. ex ante safety regulation: Substitutes or complements? *American*

Economic Review, 80, 888-901.

Lei, X.T. (2010). On assets specificity of China's enterprises: evidence from the empirical study of China's manufacturing listed companies. *Journal of Zhongnan University of Economics and Law*, (1), 101-106.

Li, J.Z., Zhou, D.Y., & Liu, C.L. (2004). Disscusion on foreign food supply chain and food safety and quality. *Foreign Economies and Management*, 26(12), 30-34.

Maldonado, E., Henson, S., Caswell, J., Leos, L., Martinez, P., Aranda, G. & Cadena, J. (2005). Cost-benefit analysis of HACCP implementation in the Mexican meat industry. *Food Control*, 16(4), 375-381.

Meuwissen, M.P.M., Velthuis, A.G.J., Hogeveen, H. & Huirne, R.B.M. (2003). Traceability and certification in meat supply chains. *Journal of Agribusiness*, 21(2), 167-182.

Ritson, C. & Mai, L.W. (1998). The economics of food safety. *Nutrition & Food Science*, 98(5), 253-259.

Souza-Monteiro, D.M. & Caswell, J.A. (2004). The economics of implementing traceability in beef supply chains: Trends in major producing and trading countries. University of Massachusetts, Amherst Working Paper No. 2004-6. Available at SSRN: http://ssrn.com/abstract=560067 or http://dx.doi.org/10.2139/ssrn.560067.

Ye, Y., Zhang, Y.H., Yue, Y. & Luo, H.E. (2011). Traceability study of food safety issues. *Journal of Huangzhong Agricultural University (Social Science Edition)*, (2), 4.

Zhou, J. & Jiang, L. (2007). An analysis on vegetable farmers' behaviors and the food safety system. *Journal of Zhejiang University (Humanities and Social Science)*, 37(2), 119-127.

Zhou, J. & Ye, J. (2007). Current situation, bottleneck and path selection for the HACCP application in food safety management in China: Analysis of agricultural product processing enterprises. *Issues in Agricultural Economy*, 8, 55-61.

Quality Perception, Safer Behavior Management and Control of Aquaculture: Experience of Exporting Enterprises of Zhejiang Province, China

8.1 Introduction

China is the biggest producer and trader of aquatic products worldwide. In 2010, the amount of aquatic products reached 53.73 million tons, up by 4.1% year on year, and exports were around 13.8 billion US dollars, ranking China first in bulk agricultural exports for 11 consecutive years.

Zhejiang Province located in the eastern region of China, the leading aquatic products exporting province in China, accounting for 10%–15% of total national exports. Its aqua-products yield about 5.30 million tons and exports reached 0.457 million tons, ranking this province fifth across China in 2011. Despite continuous growth of aquaculture production and exports in Zhejiang Province and China, increased trade frictions have caused many rejections of exported aquatic products. It thus threatens the sustainable development of China's aquaculture industry. For example, there was a sharp drop in exports in the years 2005 and 2007.

One big problem behind the recent years of blocked Chinese exports, is on account of quality and safety issues, such as exceeding the limits on fishery drug residues, disease-causing bacteria detection and so forth. What causes these issues? That is what we would like to cover based on the internal and external factors. We first list some categories of internal factors as follows: (1) problems in the environment of origin caused by water pollution, overuse of feedstuffs, and abuse of fishery drugs; (2) Inter-bureau coordination has been weak as multiple agencies at different levels have set setting standards according to their special interests, and there has been a lack of updated quality standards, in addition there has been insufficient harmonization with internationally accepted standards; (3) Quality and safety control systems are imperfect in exporting enterprises, in particular with no satisfactory adoption of self-inspection and control systems, which causes problems beyond the timely monitoring of aquatic quality and safety. The external factors focus on aquatic products export trade barriers in terms of strict trade rules and regulations. The importing countries are increasingly taking steps to enhance the use of aqua-product standards, not only to protect the domestic aqua-food market, but also to ensure food safety and quality from other countries.

Exporting enterprises could set up and improve the self-inspection system by monitoring safe production and operations, and by testing the products' quality and safety, in order to ensure that each batch of aquatic products comply with exporting standards, despite the inevitable trade barriers. In 2006, the *Agricultural Product Quality Safety Law of the People's Republic of China* was promulgated by the Chinese government, wherein it establishes a specific requirement of self-inspection at various intervals and the making of records by obtaining documentation from food enterprises. However, the limitations of small scale enterprises, lack of social and management responsibility of food processors in China, along with a combination of a wide absence of quality supervision, mean that the Chinese aquatic export enterprises haven't established a realistic self-testing system as of now.

In other words, this research has a major goal of studying factors that influence the aquatic enterprises process of routine self-inspection, and to then also explore the difficulties in adopting this system. We offer strategic insights into workable, scientific and practical policy options for strengthening China's capacity to implement standards and technical regulations for regional and international trade in aquaculture. Beneficial constructive recommendations from this research could also greatly help other developing countries to establish and implement self-inspection systems, in order to jointly improve global aquatic product quality levels.

8.2 Aquatic Products Export Restrictions: the Situation from Zhejiang Province

The export of aqua-products from Zhejiang Province is mainly focused on some developed countries and areas like Japan, Republic of Korea, the EU, ASEAN, U.S. and Canada. In 2010, the amount of export trade to the first four of those regions accounted for over 70% of trade, and of those Japan and Republic of Korea are the two biggest export markets. From figures shown in Table 8.1, we find the current situation of aquatic product exports both from Zhejiang Province and nationwide under development. We then expand our analysis to explanations of rejection issues from Japan and the United States and go deeper into the facts causing export restrictions.

Table 8.1 Rejection issues from Japan and the United States

County	Issue	Detailed information	Batch
U.S.	Quality problem	Two samples containing a putrid substance Two samples containing a filthy substance	4
	Additive problems	One sample of an unsafe food additive	1
Japan	Quality problem	Unqualified standards; Sanitary standards absent	3
	Additive problems	Additive use beyond the allowed limits	2
	Unqualified residue standards	Chemical residues exceeding standards	5
	Microbial contamination	*E. coli* and bacteria are exceeding standards and limitation)	5
	Biotoxins	Level of DSP exceeding the maximum limit	1

Data source: http://www.tbt-sps.gov.cn

8.2.1 Residues of an Agricultural Compound in Aquatic Products

There are two technology systems that impede Chinese aqua-product exports, SPS (Sanitary and Phytosanitary) and the agreement on TBT (Technical Barriers to Trade) which have been set up for the export of products. In detail, residues of agricultural products or compounds in products must be in line with SPS standards. While referring to TBT, it is closely linked to the agreement on the application of sanitary and phytosanitary measures including the requirements on labels and packaging, qualification procedures and certification regulations for agricultural products or food.

In order to find the reasons for problems that existed in recent aqua-product exports from Zhejiang Province, we will provide a summary of quality testing and control along the supply chain from production, processing to the distribution stages.

8.2.2 Safety and Quality Problems in the Aquatic Products Supply Chain

Given that Zhejiang Province agricultural exports depend largely on aquaculture and that small enterprises are the driving force behind aquatic production, there is wide attention paid to the importance of addressing aquaculture policies and especially trade-related constraints which prevent small enterprises and traders from accessing high value regional and global markets. However, some enterprises and traders still use illegal drugs and additives contrary to the standards requirements of importing countries or regions, together with an undeveloped cold chain management, though more attention is being paid to strengthening the quality and safety management of aquatic products in Zhejiang Province. All of the above indicates the relatively high quality risks which can occur during the three stages of the aquatic supply chain.

8.2.2.1 Quality Problems in the Breeding Stage

Fish farming, as the primary stage of the supply chain, has a direct impact on the aquatic product safety. There are two problems that exist in this stage to ensure product quality and safety in line with laws and regulations, to meet consumer demand and to safeguard the agricultural environment.

(1) Environmental taints

An aquaculture farm can be built on land or be set up off an ocean shore. In Zhejiang Province, nearly a half of the aquaculture harvest comes from offshore in the ocean, of which about 57% of this total is largely and seriously polluted by inorganic nitrogen, reactive phosphate and petroleum, as a result of eutrophication and red tide, according to the 2009 Environmental Report of Zhejiang Province Ocean Area. Another environmental problem comes from farmers' aiming to increase the yield of production regardless of farm density with overfeeding, overuse and abuse of drugs, thus threatening the environment and culture of the water quality.

(2) Safety and quality problems in aquatic culture

● *Safety control of aquatic fingerlings*

Over the years, due to insufficient attention to the development of fish breeding, production and management, low investment, and lack of supervision, there can be no guarantee of the quality of aquatic fingerlings production, leading to a disruption in market development and promotion. At present, China's germplasm degradation is serious and has a low coverage within good aquatic fingerlings. In addition, the insufficient number of aquatic breeding farms in China affects the quality of the seed breeding capabilities, and cannot meet the needs of the export quality of farmed species.

● *Aquaculture applied material management*

The aquaculture-applied materials include young fish, feed, feed additives, fishery drugs and other chemical compounds and biological agents (Situ, 2009). Major problems behind aquaculture-applied materials consist of low quality and abuse of applied materials. These materials indirectly or directly applied in breeding have an influence on a good aqua-culture environment and water quality, which are the prerequisites for edible safety of aquatic products.

Feed use: There are two kinds of feed for aquatic farming, natural feed and compound feed. These feeds consist entirely of raw materials and provide the feed for aquatic farming. If not inspected and not certified by some authority, aquatic product safety cannot be ensured. The aquatic compound feeds are feedstuffs that are blended from various raw materials and additives according to the specific requirements of nutrition. Potential quality accidents first come from unsuitable raw materials. Then, if the feed is overused with excessive additives fed to the fish, this may lead to an excess in additive requirements when the aquatic products are harvested.

Use of fish drugs: Some fish farms often use fishery drugs which might contain hormones and hypnotics to prevent diseases in the aquaculture. Others use drugs according to their experience and the harvest. They do not comply with the withdrawal time standards, and then cause drug residues when the aquatic product is harvested and sold to markets.

● *Input abuse in fish farming*

Fish farming has opened the path to an intensive culture with an increasing use of feed, which has contaminated the water and the environment in recent years. Other approaches aimed at protecting the environment from agriculture and industrial production can also taint the water for fish farming.

8.2.2.2 Quality and Safety in Aquatic Product Processing

Aquatic safety problems arise from food additives and abuse from illegal additives. Some aquatic enterprises do not comply with the quality and safety standards systems such as the HACCP system when they are in operating production, and this can cause potential safety issues in aquatic products.

New technologies and new materials that are used in aquatic production, for instance GM, radioactive technology, need further proof of their safety, when used in food production.

8.2.2.3 Quality and Safety at the Distribution Stage

The substandard conditions of aquatic products storage, shelflife and transportation can make quality and products unsafe. A poorly organized distribution and relatively weak aquatic product supply chain for aquaculture in Zhejiang Province prevented aquatic enterprises from implementing quality and safety traceability systems that ensure product quality and safety along the supply chain through information tracking and tracing back from the "farm to table".

8.3 Related References

Concerning studies of aquatic products based on quality and safety theory, information asymmetry and adverse selections are widely applied to explain the underlying factors. The existing references mainly focus on current development conditions in the aquatic market, construction of quality supervision, and establishment of management institutions, by taking producers and consumers as their research subjects. To date, much research has focused on the overall condition of aquaculture (Shen, 2011; Rong *et al.*, 2010; Sun *et al.*,

2007; Wu *et al.*, 2010), and others have contributed to the quality standards of farmers and enterprises (Mao, 2011; Wang *et al.*, 2011). The remaining few have concentrated on export aquatic enterprises (Cai *et al.*, 2011; Shi *et al.*, 2007; Zhang *et al.*, 2010). Due to the diverse characteristics of products, scale, and capital ability of enterprises, and also difficulties in obtaining good data from these enterprises, few studies have been conducted on enterprises' perception of quality, and their reaction to regulation and orders from the government. In particular, enterprises' safety behavior such as self-inspection, quality management and control are merely concerns.

Overall, the objective of the research is to examine enterprises' perceptions of product quality and safety, self-inspection behavior and safety behavior based on the data surveyed from aqua-culture exporting enterprises in Zhejiang Province, China. This research may also provide some contributions to the exploration of underlying factors that affect product quality and safety in aquaculture, and provide some scientific evidence for quality and safety improvement in export aquatic enterprises.

8.4 Materials and Methods

8.4.1 Data Collection

Zhejiang Province is one of the few provinces that have led the way in exploring the theories and practices related to the construction of an international level of aquatic product standards. It was also the first province to propose HACCP systems adoption in fish farming, and then also developed the first specialized HACCP plan worldwide for promotion of a health aquaculture model and operation. The seafood production in Zhejiang Province accounts for 79.8% of the total provincial aquatic production. This work focused on seafood firms including self-exportation and via export-agencies for four off-shore cities in Zhejiang Province, i.e. Ningbo, Zhoushan, Wenzhou and Taizhou. The survey conducted in the four cities is considered to be representative, because the total seafood production of these four cities makes up nearly 99.76% and 99.79% in 2009 and 2010 of provincial seafood production, respectively. Seventy firms were randomly selected throughout these cities, and we then interviewed managers from these exporting firms from May to August of 2011. Of the total firms surveyed, four of the questionnaires were considered invalid because of extensive incomplete information, and sixty-four were considered valid.

8.4.2 Methods

This research focused on self-inspection systems and especially on periodic quality control activities as major objectives with no uniform index to measure each aspect of the inspection system such as environment, equipment, employees, materials and products. Self-inspection should be conducted for monitoring the implementation and compliance with good breeding practices and for taking necessary corrective measures, when appropriate. A firm has two alternatives: to conduct or not to conduct its own security reviews. In the following discussion, we would like to examine self-inspection behavior, enterprises' perception of product quality, and other behavior related to quality management and control.

8.5 Descriptive Analysis

8.5.1 Characteristics of Sample Firms

8.5.1.1 Scale of Firms

With regard to aquatic firm size (see Table 8.2), 52.3% of the firms had annual sales below five million yuan (nearly US $786,757), 29.7% had annual sales from five to ten million yuan, and only one firm had annual sales above thirty million yuan (nearly US $4,720,543).

Table 8.2 Scale of firms, export approach and self-inspection behavior

	Firm scale (ten thousand)				Export approach	
	<500	500–1000	1000–3000	>3000	Export right	Agent export
Self-in-spection	23 (23.9%)	15 (23.4%)	8 (12.5%)	1 (1.6%)	10 (15.6%)	37 (57.8%)
Non self-in-spection	13 (20.3%)	4 (6.3%)	0	0	2 (3.1%)	15 (23.4%)
Total	36 (56.3%)	19 (29.7%)	8 (12.5%)	1 (1.6%)	12 (18.8%)	52 (81.35%)

8.5.1.2 Enterprises Conducting Self-inspection

Self-inspection is a pre-arranged program to ensure aquatic product quality in this study. Personnel matters, premises, equipment, documentation, production, quality control, distribution of the medicinal products, arrangements for dealing with complaints and recalls should all be examined at periodic intervals by designated competent persons from the company or independent audits by external experts should be conducted. 36.2% of the enterprises sampled failed to implement a self-testing process. 63.8% implemented self-inspection behavior, 10 enterprises built testing labs while the other 37 conducted self-inspection through sending their product samples to an independent audit organization. Aquatic exporting enterprises were more inclined to choose a third party to verify their product quality assurance, for they were often limited by a lack of funding and competent testing professionals.

Referring to the test items, fishery drug residuals were most common in routine testing of 47 enterprises, accounting for 73.4%, followed by heavy metals in 40 enterprises, accounting for 62.5%. Since drug residues were the main cause of aquatic export bans in Zhejiang Province, it should be noted that all enterprises with self-testing procedures conducted this test item.

8.5.1.3 Aquatic Enterprises Export Approach

About twelve firms had direct export rights, and the other 52 firms exported their aquatic products via a professional agent (Table 8.2). Firms with export rights were a smaller portion of all respondents having lengthy approval processes, more stringent quality requirements and m e difficulties in contacts with foreign inspection and testing organizations.

8.5.1.4 Product Certification

In China, there are three categories of product certifications, i.e. pollution-free products, green products and organic products. From our investigation, the total certification rate accounted for 71.9%, with pollution-free product certificates and green product certificates accounting for 42.2% (Table 8.3).

Table 8.3 Level of quality certification and product certification

	Certification level	Number	Share (%)
Quality system certification	No certification	45	70.3
	GAP	15	23.4
	HACCP	4	6.25
Product certification	No certification	10	15.6
	Pollution-free product	27	42.2
	Green product	27	42.2

8.5.1.5 Quality System Certification

There were four firms certified by the HACCP system, accounting for 6.25%, with annual sales above ten million yuan (Table 8.3). About 21.9% of respondents applied for GAP certifications. The certification rate was relatively low in fish farming enterprises, while the quality certificates were commonly available in aquatic product processing industries. This result indicated that there is a low level of adoption of the HACCP and GAP systems by fish breeding enterprises.

8.5.1.6 Main Export Markets

There are 32 enterprises (56.2%) exporting to Republic of Korea, and 16 enterprises (23.4%) exporting to Japan, and just 8 enterprises (12.5%) and 4 enterprises (6.3%) exporting to the EU and the USA, respectively. The investigation objective was oriented to marine aquaculture export enterprises, taking primary aquatic products as the main component, because that is compatible with the eating habits in South Korea and Japan, where people consume fresh and primary aquatic products. On the opposite side, America and the EU prefer processed aquatic products and the trade barriers in America and the EU are more demanding compared to those in South Korea and Japan. The EU requires that seafood entering the EU be produced according to the HACCP system of production. In the United States, HACCP systems implementation is mandatory for aquatic products.

8.5.1.7 *Export Market Characteristics*

The influence of export market characteristics on the trade in enterprises' aquatic products shows that 37 enterprises (57.8%) placed the emphasis on export competition, 60 enterprises (93.8%) on export market trade barriers, 46 enterprises (71.9%) on export market uncertainty, 29 enterprises (45.3%) on the influence of the export exchange rate and 21 enterprises (32.8%) on the distances to the export markets. As can be seen from the graph, 93.8% of the enterprises think export market trade barriers on their export trade is influential, which has become a big obstacle for aqua-food exports from Zhejiang Province. Before 2002, respondents from aquatic firms mainly had a negative response when facing trade barriers. However, after 2002, with the increasing requirements on exported aquatic products by the importing countries, the government and industries from all walks of life paid more attention to issues of trade barriers. The enterprises themselves have also actively taken measures to deal with the export trade restrictions, such as traceability system implementation and self-inspection setups.

62.5% of the enterprises agreed that export market uncertainty, such as the financial crisis, the Japanese tsunami and other factors, had direct effects on aquatic product exports. In 2011, due to the tsunami and the nuclear leak in Japan, Zhejiang Province seafood export orders to Republic of Korea greatly increased. Export market uncertainty represents obscure threats that are difficult for domestic enterprises to predict. Sharpening product quality levels and increasing their competitive power will be their only way to maintain aquatic product exports.

8.5.1.8 The Influence of Enterprises' Export Trade Barriers

Thirty-nine enterprises out of all the respondents interviewed had experienced one or more export bans, accounting for 68.8% of the total sample. The export bans are increasingly becoming a big obstacle for Zhejiang Province aquatic products export trade. The main reason for these export bans was on account of the drug residues exceeding standards, accounting for 41.2% of the total, followed by heavy metals and microbe residuals. Enterprises should earnestly address aquatic products quality and safety, and real-time monitoring of aquatic products residues, in order to avoid export bans. Packaging and labeling requirements in recent years have become more and more strict, especially for green packaging requirements. European countries and America's requirements for traceability labels place higher requirements on Zhejiang Province enterprises when exporting aquatic products. Water from aquaculture farms that was not in line with the standards was another environmental blot on aquatic quality. The enterprises that could not meet technical requirements were the enterprises that had breeding operations integrated with processing, and the techniques applied in aquatic processing did not comply with the standards of the importing countries.

It can be shown that the reasons the surveyed enterprises suffered from export bans are as follows (see Table 8.4): International standards are too rigorous for domestic aquatic enterprises to reach in terms of current aquaculture techniques (40, 62.5%); foreign standards updates are too quick to allow follow up in a timely manner, and require the enterprises' continuous investment, which increases costs (15, 23.4%); complicated procedures for applying for foreign export licenses (38, 59.4%); strict requirements of foreign customs inspection (27, 42.2%); enterprises cannot afford the burden of increasing production management fees in order to meet strict foreign requirements (23, 35.9%).

Table 8.4 Reasons for aquatic products export bans

	No. of firms	Share (%)
Costly testing	23	35.9
Strict requirements	27	42.2
Complicated procedures	38	59.4
Fast updates of foreign standards	15	23.4
No international certification	6	9.4
Higher level of international standards	40	62.5

Because the requirements set by the state for the health and technical standards are according to the latest equipment and technical methods of determination, the enterprise's production techniques are behind those in developed countries, so enterprises in general cannot meet the requirements, unless personnel, material and financial resources to develop these new techniques and advanced equipment are put in place.

8.5.2 Quality Perception of Aquatic Exporting Enterprises

Referring to the perception of quality of feeds and feed additives, aquaculture enterprises have a basic or good knowledge of those requirements (Table 8.5). According to the Regulation on the Management of Feeds and Feed Additives, the management of feeds and feed additives aims to ensure the quality and safety of animal products as well as maintaining the health of the public. Enterprises' perceived quality compiled from fishery drugs are shown in Table 8.5. All enterprises had the knowledge that the fishery drugs applied would affect the aquatic product quality. Compared with the group on feed quality perception, fewer respondents had a good knowledge of quality perception of fishery drugs compared to that of fish feed.

All respondents surveyed complied with requirements of the withdrawal time, after which drugs may not be added. But in some cases fishery drugs were added to fishery feed which remain within the fish when caught, if added too late. For these exporting enterprises, the withdrawal times differed relative to the type of fish so as to meet the demands of the importing markets concerning the avoidance of drug residues.

Table 8.5 Enterprises' perception of aquatic product quality

		Self-inspection	Non self-inspection	Total
Feeds quality	Good knowledge	29	14	43
	Basic knowledge	18	3	21
Fishery drugs	Good knowledge	24	10	34
	Basic knowledge	23	7	30
Withdrawal period	Good knowledge	47	17	64
Production record	Important	21	5	26
	Fair important	20	7	27
	Less important	6	5	11

There were fifty-three enterprises that chose to make a production record during aquaculture for ensuring the aquatic products quality (Table 8.5). The possible reason is that by keeping detailed information of breeding, these enterprises could have real-time control on their production and safety activities. This information from production documents then provided original records for self-inspection on quality and other related testing, which could be documented as test reports and forwarded to the export organizations for testing and quarantine.

8.5.3 Safety Management and Control Behavior

Food safety behavior relates to enterprises' activities concerned with guaranteeing food safety and quality. To aquatic enterprises, these activities include fingerlings selection, feed and fishery drug control, disease examination, quality control, information documentation, surveillance and supervision, environmental protection and remedies, and relevant research development, constitution building and staffing along the whole aquaculture supply chain.

8.5.3.1 Use of Aquaculture Applied Materials

(1) Aquatic seed purchasing behavior

From the survey, there were twenty-six enterprises breeding seeds themselves. Thirty-eight enterprises purchased fish seed from other firms or aquatic farms. Among these, 28 did their procurement from specialized breeding companies and fish seed farms, 10 purchased from common traders, typically small-scale enterprises with low quality requirements for seed. Fifty-eight enterprises placing a major concern on high quality (90.6%) made up a large portion of all respondents, indicating that there were major concerns by managers from aquaculture enterprises on fish seed quality.

Feeds are the basis of aquaculture-applied materials. The costs account for almost 70% of the total costs incurred in fish farming from the surveyed respondents. Nutritional content and price of feed were the major concerns; 46.9% of enterprises applied compound feed by mixing it with natural feed, and 28.1% chose compound feed only, while the remainder chose to use natural feed only.

Twenty-four respondents used fishery drugs accounting for 37.5% of the total respondents and 40 enterprises did not use fishery drugs at all. Of these, 29 (45.3%) enterprises followed the instructions in the fishery drugs handbooks, 14 (21.9%) were under the guidance of the aquatic disease control staff and twenty-two applied for training knowledge for pesticide applications. It indicates that a high number of enterprises were in accordance with the standards. However, there were still 24 (37.5%) by experience that were not in accordance with the standards. With the increase in sea pollution in recent years, such as heavy metal pollution, and also from the possibility of new diseases being found in the waters, farming enterprises cannot just rely on their own experiences, but attention should also be paid to disease prevention and treatment via scientific methods.

(2) Aquaculture production documentation

The results show that all enterprises recorded many aspects of their aquariums except for 10 enterprises with agency exports. The contents of these records contain information about feed sources and breeding, fish diseases and fishery drug use, water quality, fingerlings source and so forth. A possible explanation is that importing countries, especially those developed countries, have stringent standards on production records from importing aquatic enterprises. In addition, requirements for traceable labeling forced these exporting enterprises to maintain breeding documentation.

8.5.3.2 Greater Government Support and Expected Involvement in Aquatic Quality Supervision

The central government is actively involved in enterprises' conducting self-inspection in terms of finance and policy support, technical training and guidance, support for assistance in export certification, enforcement of governmental inspection and testing, enforcement of aquaculture regulations, information about aquatic product export markets, and training on techniques of importing countries. The main goal of such measures is to ultimately incentivize enterprises to conduct self-testing. The results indicated that 76.6% of enterprises hoped that greater government involvement would offer financial support, 73.9% of enterprises hoped national and local governments' policy support, and a third hoped that support would come for assistance with export certifications.

8.5.3.3 Industry Associations for Aquaculture

Aquatic industry associations have been spawned over recent years by government agencies that have begun turning to these associations for help in shaping technical standards. In our investigation, fifty-four firms have joined in aquatic association accounting. Fifty-four enterprises hoped that support would come from industry associations related to building testing labs for conducting self-inspection by themselves. 49 firms hoped support would come for assistance in preparing exporting certifications. In addition, 38 enterprises were looking for further assistance on export markets promotion and technique guidance and an inquiry service for industry associations.

These industry associations also play an important role in activities such as fishery information issues, holding new technique training, assisting in the exploration of new sales channels, which help motivate enterprises to invest in aquaculture (Wang and Song, 2010).

8.6 Conclusions and Recommendations

Production records of aquaculture should be established to strengthen quality and safety of aquatic products and to reduce the safety risk. Repeated exposure to quality and safety issues, export detention and return of aquatic products, indicate that policy makers should strengthen the source management regarding the quality of aquatic products of enterprises in terms of carrying out the legal measures, and supervise the enterprises to implement an effective product inspection system to ensure quality and safety of aquatic products.

Construction of the quality and safety management system and product certification of aquatic products should also be promoted and improved. The results indicate that the exporting enterprises which have certification by HACCP and GAP have a better implementation of the self-inspection behavior. In general, HACCP certifications and GAP certifications are recognized as quality and safety certification systems worldwide, which can effectively control the quality and safety of aquatic products. Large-scale enterprises have the capital and special personnel to establish self-testing systems, and directly receive the inspection and quarantine of foreign entry-exit inspection and quarantine bureau. In addition, they face directly the customers in overseas markets, so they place a high value on quality and safety of aquatic products and the self-inspection systems.

In order to improve the certification information systems of aquatic products, the relevant departments should release accurate standards in a timely manner, and offer assistance to exporting enterprises and these inspection agencies should help them develop comprehensive and detailed testing standards for export products. These measures can make them keep up with constantly updated international standards and break through the bottleneck of

China's lagging testing standards.

Based on these findings in our paper, we propose several policy recommendations to promote self-inspection behavior, and aquatic product safety perception as well in China.

As a powerful promoter for enterprises to provide safe aquatic products, the support and relevant involvement of the government have positive effects on enterprises' willingness to cooperate. Such involvement through bonding between public third party testing and production and processing enterprises will contribute to aquatic enterprises' self-inspection willingness and behavior. In order to motivate aquatic self-inspection behavior, a fullproof system of periodical testing and supervision is demanded, which consists of quality management, production surveillance and regulation on market access.

Government should improve the level of administrative services. As the largest service and regulatory organization, the government should play a two-pronged role in services and supervision by carrying out more targeted technical training and guidance for aquatic enterprises, to offer more training for professional inspection staff in enterprises. More highly educated and literate personnel should be in the inspection and quarantine team. At the same time, the government should provide enterprises with more information to help them obtain export product certification, and by combining this with a better and more comprehensive understanding of the system of regular casual inspections of the company's products.

In the future, a new model that consists of flagship processing firms-led and small scale aqua-culture enterprises should be given more attention. These processing firms can exert quality management and control over aquatic farms that are owned by some individual farmers or small scale firms, that will help to improve farmers' awareness of aquatic product quality (Cai *et al.*, 2011). It therefore helps to bring a stable relationship of material supply from farms to processing enterprises. The processing enterprises will conduct a transformation of export trade development via quality value-added techniques, rather than through traditional trade in large quantities. The change will improve aquaculture quality and also their other products, and provide benefits for the aquatic market domestically and worldwide. Therefore, it promotes the development of international trade especially for these small- and middle-scale enterprises led by those flagship exporting aquatic product processors.

References

Cai, Q., Fan, Y. & Zhuang, P. (2011). On the export competitiveness of Fujian aquatic products. *Journal of Fujian Agriculture and Forestry University (Philosophy and Social Sciences)*, 14(1), 34-38 (in Chinese).

Mao, J. (2011). The sea fishery power compensation mechanism from the fisherman angle. *Chinese Fisheries Economics*, 29(06), 50-53 (in Chinese).

Rong, C.,Gou, W. & Sun, P. (2010). Cooperation among the enterprises on aquaculture industry chain management based on game theory. *Chinese Fisheries Economics*, 28(04), 116-121 (in Chinese).

Shen, Q. (2011). Study on China's current quality and safety of aquatic products and supervision measures. *Beijing Agriculture*, (4), 35-36 (in Chinese).

Shi, D., Yang, F. & Ye, Y. (2007). Structure of exporting agricultural products in Guangdong Province. *Guangdong Agricultural Sciences*, (08), 103-105 (in Chinese).

Situ, J. (2009). The effect of aquaculture applied materials on aquatic production quality. *Agricultural Aquaculture*, (22), 27-28 (in Chinese).

Sun, Z., Qian, H. & Chen, F. (2007). Policy of fishery safety and quality. *Food Science and Technology*, (02), 5-7 (in Chinese).

Wang, S. & Song, Y. (2010). Construction and perfection of fishery products quality and safety management system in China. *Chinese Fisheries Economics*, 28(1), 3-7 (in Chinese).

Wang, S., Liu, Q. & Wang, J. (2011). An empirical analysis of the quality recognition and behavior selection for aquatic farms: a case study of Shanghai. *Chinese Fisheries Economics*, 29(4), 64-70 (in Chinese).

Wu, H., Lu, F., Li, X., Liu, Q. & Wang, G. (2010). The existing problems and the corresponding upgrading strategies of large-scale aquaculture of freshwater fish circulation chain in China. *Chinese Fisheries Economics*, 28(06), 44-49 (in Chinese).

Zhang, L. & Zhang, D. (2010). Relationship between aquatic products exports and economic growth: Experience from Shenjiang. *Economic Research Guide*, (21), 145-148 (in Chinese).

Chapter

9

Outlook for China's Food Safety Situation and Policy Recommendations

9.1 Outlook for China's Food Safety Situation

Since the beginning of 2011, food safety incidents have occurred now and then, and food safety will continue to be a hot topic. The continuous exposure of China's food safety incidents shows that the public is more concerned about food safety. But it is fair to say that the overall situation of China's food quality is steadily improving, because the CPC Central Committee and the State Council attach great importance to food quality safety and governments at all levels continue to strengthen food safety administration and have provided much more support in policy and finance. But on the other hand, we must clearly recognize that compared with the increasingly strict requirements of the CPC Central Committee and the State Council for food quality safety, compared with the increasing expectations of the public for food quality safety, and compared with the level of developed countries, the actual state and level of food safety in China still lags. Under these circumstances, combined with the above analysis and summary of the status quo of China's food safety, we look forward to the future of China's food safety situation from three perspectives, namely, consumers, industry, and supervision.

(1) From the consumers' point of view, the trend of green consumption and safety consumption has emerged. With the upgrade of the consumption structure, the consumer requirements for food safety have changed from "having enough to eat" to "having better and finer food to eat". "Health care" has become the dominant trend of the food and beverage industry, and this trend will continue into 2012. Food with good quality and trophic structure has become the first choice for consumers. Demand for green food, organic food and other medium and high-end food will gradually increase. In addition, the development of e-commerce of agricultural products and the link-up of farmers and stores have provided opportunities for the rise of green consumption.

(2) Viewed from the perspective of the food industry, the level of industry concentration will further be enhanced and the level of industry safety will be steadily improving. First, the level of concentration of the food industry will be further improved. Because, currently, prices of agricultural products in China remain at a high level, the pressure of rising production costs of the food industry is still relatively large, and in addition consumer concern for food quality safety and the strengthening of policy supervision will, to a certain extent, increase the cost of food companies. Under such circumstances, it is an ideal choice to give full play to the economies of scale of the food processing industry. Therefore, we expect that some capable food processing enterprises will expand to the upper and lower reaches of the industrial chain, and further integrate the production stage of food raw materials and combine it with the circulation stage. While ensuring the quality of raw materials, this kind of expansion also helps to reduce the stages of circulation and decrease circulation costs, thus effectively reducing operating costs. Meanwhile, the expansion of enterprises can also contribute to the improvement of the concentration of the entire food industry, and help enhance the competitiveness of the industry as a whole. Second, apart from the expansion in scale, the over-

all safety level of the food industry will steadily increase. The enhancement of the level of food safety results from not only the policy but also from the enterprises' own initiative to meet the consumption trends. In 2011, China carried out a series of special operations against illegal behavior of the whole food industry, such as Special Operation against Illegal Food Additives and Special Operation against Lean Meat Powder, which shows the country's determination to maintain food safety. Due to these policies, food industry enterprises had to ensure the quality of their products. In addition, the coming of the green consumption trend and the increasing cost of materials also forced food enterprises to gradually change their marketing strategy from "Assuring success through competitive pricing, Small profit and quick return" to entering medium and high-end food markets.

(3) Viewing from the perspective of a food safety supervision policy, we predict that food safety early warning, capacity of handling major events, and the establishment of traceability systems will become the focus of government policy. Meanwhile, government investment in the wholesale markets of agricultural products and cold chain logistics infrastructure will also be gradually increased.

First, the occurrence of "Lean Meat Powder" incident, "Gutter Oil" incident, and other major food safety incidents shows that how to prevent such incidents and how to enhance the capacity of the government and enterprises to address major emergencies has become the key to deal with food safety incidents in the future. Following the issuance of Law of the People's Republic of China on Emergency Responses, Food Safety Law of the People's Republic of China, Law of the People's Republic of China on Agricultural Product Quality Safety, Enforcement Regulations of Food Safety Law of the People's Republic of China, Regulations on the Preparedness for and Response to Emergent Public Health Hazards, and Overall Contingency Plan for National Public Emergencies, in October 2011 the State Council amended the Contingency Plan for National Food Safety Incidents. A series of documents and files reflect that the central government will focus on strengthening the preliminary risk assessment, public psychological intervention, and risk communication of major food safety incidents, so as to enhance the capacity of addressing emergencies.

Second, with the issuance and implementation of Instructions of Ministry of Commerce on the Construction of a Traceability System for Meat and Vegetable Circulation in the Twelfth Five-Year Plan Period, the Ministry of Commerce and the Ministry of Finance chose, in two installments beginning in 2010, 20 large and medium-sized pilot cities across the country to construct traceability systems for meat and vegetable circulation. Meanwhile, the pace of constructing city traceability systems is also accelerating. In 2012, the government will gradually expand the traceability system to cattle, sheep, chickens, ducks and other poultry products, fruits, aquatic products, edible mushrooms, soybean products, and other major products in all cities across the country that have a population over one million.

Third, the infrastructure construction of food distribution will also be an important work of the government next year. As the core stage of the distribution of agricultural products, the agricultural product wholesale market is of great significance to food safety. But the current operational system for wholesale markets, to a large extent, restricts the role of wholesale markets in supervising food safety, so that it is a viable option to establish non-profit wholesale markets. What's more, the construction of wholesale markets is significant to crack the current dilemma "difficult to sell agricultural products". Beginning in 2010, whether it is the state's direct financial investment or a variety of preferential policies, all these efforts confirm that the government attaches much importance to the construction of wholesale markets. In addition, a serious shortage of cold chain and cold storage is one of the key factors for the high cost of the distribution of agricultural products. Development Plan for Agricultural Product Cold Chain Logistics prescribes that by 2015, the cold chain distribution rate of fruits and vegetables, meat, and aquatic products will increase to over 20%, 30%, and 36% respectively, the refrigerated transport rate will increase to about 30%, 50%, and 65% respectively, and the decay rate of agricultural products in distribution stage will decrease to 15%, 8%, and 10% respectively. To achieve this goal, within the next few years, the government must construct large-scale cross-region cold chain logistics distribution centers that are of high efficiency and apply new technologies, and cultivate core cold chain logistics enterprises that have strong ability to integrate resources and have strong international competitiveness, and tentatively build a service system for agricultural product cold chain logistics.

9.2 Policy Recommendations

In fact, in recent years, China has taken a number of strong measures to strengthen the monitoring of food quality safety, but food safety incidents still occur. The recurrence and intractability of incidents provides us further food for thought: apart from the reasons that the supervision system is imperfect and a small number of unscrupulous manufacturers forget their integrity when tempted by personal gain, we should investigate the deep reasons why it is difficult to carry out food safety supervision, such as price and cost, decentralized production and segmented supervision, and the standards of food quality. Therefore, to improve the food quality safety supervision within our country, we should mainly carry out the following work in the future:

9.2.1 Vigorously Develop Agricultural Cooperatives and Raise the Level of Organization in Agricultural Production and Marketing

On the one hand, the capacity of food safety supervision in China's rural area is low, and the coverage is limited. On the other hand, the level of organization in agricultural producers in China is low, and there are a huge number of agricultural producers and they are extremely scattered, resulting in high costs of food safety supervision of agricultural production processes. The improvement of food safety supervision in rural areas is a gradual process, and it cannot be accomplished in a short term. Therefore, it is an optimal choice to ensure that the limited resources of rural food safety supervision are most effectively used. Among them, raising the level of the organization in agricultural production and marketing and vigorously developing agricultural cooperatives are effective measures to improve the efficiency of supervision and enhance the rural food safety level. Studies have shown that agricultural cooperatives have a distinct advantage in improving the level of quality safety of agricultural products. First, it helps supervisiory bodies to reduce supervision costs, expand the scope of supervision, and improve the supervision effect. After the establishment of agricultural cooperatives, food safety supervisiory bodies can shift their work from scattered farmers as in the past to cooperatives. In this way, they can achieve a wider range of indirect control over cooperatives, reduce the number of direct supervision objectives, and thus reduce supervision costs. In addition, the reduction of direct supervision objectives means that the supervisiory bodies can invest more resources to each supervision objective, thereby contributing to the deepening of the supervision work. Second, the establishment of agricultural cooperatives can help agricultural producers reduce the cost of carrying out food safety work. Gathering scattered household production can play on the scale advantages of cooperatives in the procurement of production materials, product centralized detection, and technology services, saving costs of food safety work for farmers in cooperatives. Third, the establishment of agricultural cooperatives also provides an effective platform for the promotion of food safety technology and standards.

9.2.2 Strengthen Supervision of Small Food Processing Enterprises

Empirical data shows that most food safety incidents happen at the food processing stage, which is of course due to the complex technological processes, complex procedures, low threshold, and large quantity of enterprises in the food processing industry, but more importantly, because our country is lacking in supervision of small food processing enterprises that have high potential safety hazards.

Carrying out more special programs to address problems of small food processing enterprises has an immediate impact on solving food safety incidents.

In this regard, we should do the following: (1) Focus on the key points. Focus on key problems, including the use of non-food raw materials in food production, the use of food additives, "three-no-food (It refers to food with no date of production, no producer, no certification of fitness.)", and dirty, chaotic, and poor food production environments. Focus on key enterprises, including enterprises without production licenses and enterprises without production licenses and business licenses. Focus on key areas, including rural areas, urban fringe, and especially those towns and villages and urban fringe areas that are subject to food safety problems or other related potential problems. (2) Strengthen implementation and be strict. Rigorously crack down on enterprises without production licenses, resolutely ban enterprises without production licenses and business licenses, and increase the punishment for illegal processing behaviors. (3) Establish comprehensive enterprise records. Increase random inspection of processing enterprises having potential food safety problems, and carry out surprise checks. (4) Perfect the process for information release. Release the list of illegal enterprises and substandard food production to the public through the news media in a timely manner. (5) Encourage and coordinate the action of everyone and every institution. Take effective measures to encourage consumers to expose the illegal behavior of small food processing enterprises, and promote the formation of an environment in which the whole society attaches importance to food safety.

9.2.3 Establish a Number of Non-Profit Wholesale Food Markets and Strengthen the Supervision of Food Safety in Wholesale Markets

Studies have shown that the departure from or absence of a public nature for agricultural products wholesale markets is the basic reason why agricultural products wholesale markets cannot play a role in food quality safety supervision, and it also contributes to the irrational rise of prices for agricultural products. The departure from a public nature is closely related to the current policies and modes of wholesale market operations. For the past 20 years, China has been implementing a "who invests, who benefits" policy and the wholesale markets are mainly established by enterprises, so the operational mode also refers to that of joint-stock companies whose main goal is the pursuit of profits. That's why wholesale market operators often, for their own sake, charge too much pitch fees and admission fees, etc. but ignore food safety testing and reduce sampling frequency and the number of sampling items, which in turn constrains the role of wholesale markets. Therefore, wholesale food markets must return to a public nature.

9.2.4 Improve the System of Food Safety Standards and Regulate Food Quality Safety Certifications

China has established a relatively complete system of food quality safety standards that covers extensive fields and has a relatively reasonable structure. However, the overall level of the standards is still low, standards at different levels are often repetitious and inconsistent, some important standards for new technology, new energy, and GM products are inadequate, some standards are not exercisable, and some standards fail to be effectively implemented. These problems largely restrict the work of food safety supervisory departments as well as the operation of food producers. Thus, during the period ahead, we must further improve the construction of China's food safety standard systems: First, further clean-up of current food standards and solving the problems that some standards are repetitious and inconsistent is needed. Second, strengthen the basic research on standards, enhance the overall level and operability of standards, accelerate the pace of amending standards, and link up with international standards and foreign advanced standards. Third, strengthen related publication, training, implementation, and supervision.

As for food certification, the Certification and Accreditation Administration of the People's Republic of China manages, supervises, and comprehensively coordinates the certification and accreditation work. Currently, a certification and accreditation work situation, which is characterized with unified management, standardized operation and coordinated implementation, has been basically formed; a certification and accreditation system, which covers the whole production process of food and agricultural products, i.e. "from farm to fork", has been basically set up. The certification items include feed product certification, certification of good agricultural practice (GAP), certification of pollution-free agricultural products, certification of organic products, food quality certification, certification of HACCP management systems, and green market certification, etc. However, China's food safety certification management still has a series of problems: the certification system is imperfect, food certification consulting and training institutions are scarce, professional skills and talents are inadequate, food certification methods and certification standards are not in line with international standards, and third-party certification management is chaotic. So in the coming period, we must improve the food safety standards system, strengthen international cooperation in food safety certification, and rectify third party certification institutions, so as to push forward China's food safety certification work.

9.2.5 Strictly Implement Food Quality Safety Market Access Systems

Early in 2001, China had already established food quality safety market access systems, which included production licensing systems, mandatory inspection systems, and market access labeling systems. At present, China has promoted and implemented market access systems for 28 categories of 525 different kinds of food products. Although China's food quality safety market access system developed very quickly, there are still a lot of problems: the proportion of certified companies is relatively low, consumer acceptance of market access labels is not high, and the degree of information available is relatively low, which constraints the role of market access systems on ensuring food quality safety. Therefore, in future revisions, in order to improve the system, a unified food safety information system and an information platform for quality safety market access and supervision must be established, and efforts to promote and publicize this system must be strengthened; meanwhile, food quality safety market access systems should be gradually and fully put into practice through the improvement of a food safety supervision system.

In short, it is a huge project to improve China's food safety supervision system. According to the experience of developed countries, an effective food safety supervision system should include a unified national legal framework, an effective food control management system, a food safety standards system that is in line with international standards on the basis of risk prevention, a unified, authoritative, and effective food safety inspection and acceptance system, a unified and standardized food certification and accreditation system, an effective food safety emergency response system, an effective food quality safety traceability system, a national food-borne disease surveillance system, an information services system that can effectively link the supervisory bodies, consumers, and the news media, effective food safety education and training, and a variety of international food safety cooperation and technical exchanges. We should realize the persistence of China's food safety work, based on national conditions, learn from international experience, and gradually and in an orderly manner push forward the construction of China's food safety supervision system.

Food Safety Law of the People's Republic of China

(Adopted at the 7th Session of the Standing Committee of the 11th National People's Congress on February 28, 2009)

Contents

Chapter I General Provisions

Article 1 This Law is enacted to ensure the food safety and guarantee the safety of the lives and health of the general public.

Article 2 Those engaging in the following activities within the People's Republic of China shall abide by this Law:

1. Food production and processing (hereinafter referred to as food production), and food circulation and catering services (hereinafter referred to as food business operation);

2. The production and business operation of food additives;

3. The production and business operation of packing materials, containers, detergents and disinfectants for food and utensils and equipment for food production and business operation (hereinafter referred to as "food-related products");

4. The use of food additives and food-related products by food producers and business operators; and

5. The safety management of food, food additives and food-related products.

The quality and safety management of edible primary products sourced from agriculture (hereinafter referred to as "edible agricultural products") shall be governed by the provisions of the Law on the Quality and Safety of Agricultural Products. However, the formulation of quality and safety standards for edible agricultural products and the release of safety information about edible agricultural products shall be governed by the relevant provisions of this Law.

Article 3 Food producers and business operators shall follow relevant laws, regulations and food safety standards when engaging in food production and business operation activities, be responsible to the society and the general public, ensure food safety, accept social supervision and assume social responsibilities.

Article 4 The State Council shall establish a Food Safety Committee, of which the functions shall be prescribed by the State Council.

The health administrative department of the State Council shall undertake the comprehensive coordination function for food safety, be responsible for the assessment of food safety risks, formulation of food safety standards, release of food safety information, formulation of qualification determination conditions and inspection requirements for food inspection agencies, and organize the investigation and handling of major food safety accidents.

The quality supervision department, industry and commerce administrative department and state food and drug administrative department of the State Council shall, according to the functions as prescribed in this Law and those as provided for by the State Council, supervise and administer the food production, food circulation and catering services, respectively.

Article 5 A local people's government at or above the county level shall undertake the overall responsibility for the food safety supervision and administration within its own administrative region, uniformly lead, organize and coordinate the work of food safety supervision and administration within its own administrative region, establish a sound whole-process food safety supervision and administration mechanism, uniformly lead and exercise command in response to food safety emergencies, improve and execute the food safety supervision and administration accountability system, and appraise, discuss and evaluate the performances of the food safety supervision and administration departments.

A local people's government at or above the county level shall, in accordance with this Law and the provisions of the State Council, determine the food safety supervision and administration functions of the health administrative department, agriculture administrative department, quality supervision department, industry and commerce administrative department, food and drug supervision and administration department at the same level. These departments shall, within the scope of their respective functions, be responsible for the food safety supervision and administration within that administrative region.

The agency established within an administrative region at a lower level by a department of the people's government at a higher level shall do a good job in the food safety supervision and administration under the uniform coordination of the local people's government.

Article 6 The health administrative departments, agriculture administrative departments, quality supervision departments, industry and commerce administrative departments, food and drug supervision and administration departments at and above the county level shall strengthen communication and closely cooperate with each other, and exercise the powers and assume the responsibilities under their respective functions.

Article 7 The relevant food industry associations shall strengthen the industrial self-discipline, direct food producers and business operators to engage in production and business operation according to law, boost the industrial trustworthiness, publicize and popularize the knowledge on food safety.

Article 8 The state shall encourage social groups and autonomous grass-roots mass organizations to carry out the work in respect of the popularization of food safety laws, regulations, standards and knowledge, advocate healthy eating styles, and enhance consumers' food safety awareness and self-protection capability.

The news media shall publicize food safety laws, regulations, standards and knowledge for the public good and, through public opinions, supervise violations of this Law.

Article 9 The state shall encourage and support the basic research and application research relevant to food safety, and encourage food producers and business operators to adopt advanced technologies and advanced management criterions and grant support to them so as to enhance food safety levels.

Article 10 Any entity or individual shall be entitled to report any violation of this Law which is committed during the food production and business operation process, get food safety information from relevant departments and put forward opinions and suggestions on the food safety supervision and administration work.

Chapter II Monitoring and Assessment of Food Safety Risks

Article 11 The state shall establish a food safety risk monitoring system to monitor the food-borne diseases, food contamination and harmful factors in food.

The health administrative department of the State Council shall, jointly with relevant departments of the State Council, work out and execute the national food safety risk monitoring plan. The health administrative departments of the people's governments of the provinces, autonomous regions and municipalities directly under the Central Government shall, according to the national food safety risk monitoring plan, and by taking into account the actualities of their respective administration region, organize the preparation and execution of the food safety risk monitoring program for their respective administrative region.

Article 12 After getting the information about the relevant food safety risks, the agriculture administrative department, quality supervision department, industry and commerce administrative department, state food and drug administrative department and other relevant departments of the State Council shall promptly notify the health administrative department of the State Council. After verifying the information jointly with the relevant departments, the health administrative department shall timely adjust the food safety risk monitoring plan.

Article 13 The state shall establish a food safety risk assessment system to conduct risk assessment on the biological, chemical and physical hazards in food and food additives.

The health administrative department of the State Council shall be responsible for organizing the food safety risk assessment work. It shall form a food safety risk assessment expert committee composed of experts in medical science, agriculture, food, nutrition, etc., to assess food safety risks.

The safety assessment of pesticides, fertilizers, growth regulators, veterinary medicines, feeds and feed additives, etc. shall be made with the participation of experts from the food safety risk assessment expert committee.

The food safety risk assessment shall be made through scientific methods and be based on the food safety risk monitoring information, scientific data and other relevant information.

Article 14 Where the health administrative department of the State Council finds any hidden food safety risk through food safety risk monitoring or through a tip-off it receives, it shall immediately organize an inspection and a food safety risk assessment.

Article 15 The agriculture administrative department, quality supervision department, industry and commerce administrative department, state food and drug administrative department and other relevant departments of the State Council shall put forward suggestions on food safety risk assessment and furnish relevant information and materials to the health administrative department of the State Council.

The health administrative department of the State Council shall timely notify the relevant departments of the State Council of the result of food safety risk assessment.

Article 16 The result of food safety risk assessment is the scientific basis for formulating and revising the food safety standards, and for exercising food safety supervision and administration.

If it concludes from the result of food safety risk assessment that any food is unsafe, the quality supervision department, industry and commerce administrative department and state food and drug administrative department of the State Council shall, according to their respective functions, immediately take corresponding measures to ensure cessation of the production and business operation of the food in question, and inform the consumers that they should stop eating it. If it is necessary to formulate or revise the pertinent national food safety standards, the health administrative department of the State Council shall do so promptly.

Article 17 The health administrative department of the State Council shall, jointly with the relevant departments of the State Council, make a comprehensive analysis on the status quo of food safety in light of the food safety risk assessment result and the food safety supervision and administration information. If it shows that any food is with possibly high safety risk upon the comprehensive analysis, the health administrative department of the State Council shall timely give a warning of food safety risk and make an announcement.

Chapter III Food Safety Standards

Article 18 The purpose of formulating food safety standards shall be to ensure the physical health of the general public. The food safety standards shall be scientific, reasonable, safe and reliable.

Article 19 The food safety standards are standards for mandatory execution. Except for food safety standards, no other mandatory food standards shall be set down.

Article 20 The food safety standards shall contain

1. Provisions on limits of pathogenic microorganisms, pesticide residues, veterinary medicine residues, heavy metals, pollutants and other substances hazardous to human health in food and food-related products;

2. Varieties, extent of use and dosages of food additives;

3. Nutrient content requirements for staple and supplementary food exclusively for infants and other particular groups of people;

4. Requirements for labels, marks and instructions relating to food safety or nutrition;

5. Hygienic requirements for food production or business operation process;

6. Quality requirements relating to food safety;

7. Methods and procedures for food inspection; and

8. Other contents which are necessary to be formulated as food safety standards.

Article 21 The national food safety standards shall be formulated and announced by the health administrative department of the State Council, for which the standardization administrative department of the State Council shall provide the serial number of national standards.

The provisions on limits of pesticide residues and veterinary medicine residues, and the inspection methods and procedures thereof shall be formulated by the health administrative department and agriculture administrative department of the State Council.

The inspection procedures for slaughtered livestock and poultry shall be formulated by the relevant competent department of the State Council jointly with the health administrative department of the State Council.

Where any national product standard involves provisions of the national food safety standards, it shall assure its consistency with the national food safety standards.

Article 22 The health administrative department of the State Council shall consolidate the mandatory standards in the existing edible agricultural product quality and safety standards, food safety standards, food quality standards as well as relevant industrial standards on food and uniformly publish them as national food safety standards.

Before the national food safety standards as prescribed in this Law are published, the food producers and business operators shall produce food and engage in the business operation of food under the existing edible agricultural product quality and safety standards, food safety standards, food quality standards as well as relevant industrial standards on food.

Article 23 The national food safety standards shall be examined and adopted by the National Food Safety Standard Review Committee. The National Food Safety Standard Review Committee shall consist of experts in medical science, agriculture, food, nutrition, etc. and representatives from relevant departments of the State Council.

The national food safety standards shall be formulated on the basis of the food safety risk assessment results, by taking into full consideration the quality and safety risk assessment results of edible agricultural products, referring to the relevant international standards and international food safety risk assessment results, and upon soliciting opinions from a wide range of food producers, business operators and consumers.

Article 24 In the absence of national food safety standards, local food safety standards may be formulated.

When organizing the formulation of local food safety standards, the health administrative department of the people's government of a province, autonomous region or municipality directly under the Central Government shall refer to the provisions of this Law regarding the formulation of national food safety standards and report them to the health administrative department of the State Council for archival purposes.

Article 25 In the absence of national food standards or local standards for the food produced by an enterprise, the enterprise shall formulate enterprise standards as the basis for organizing the production thereof. The state shall encourage food production enterprises to formulate standards more stringent than the national food safety standards or than the local food safety standards. The standards of an enterprise shall be submitted to the provincial health administrative department for archival purposes and be applied inside the said enterprise.

Article 26 The food safety standards shall be available for the general public to consult free of charge.

Chapter IV Food Production and Business Operation

Article 27 A food producer or business operator shall meet the food safety standards and satisfy the following requirements:

1. Having places for treating food raw materials and food processing, packaging and storage, which adapt to the varieties and quantities of the food under its production or business operation; keeping the environment of the said places tidy and clean, and ensuring that they are at a prescribed distance from toxic and hazardous sites and other pollution sources;

2. Having production or business operation equipment or facilities, which adapt to the varieties and quantities of the food under its production or business operation, and having the corresponding equipment or facilities for disinfection, changing clothes, toilet, day-lighting, illumination, ventilation, anti-corrosion, anti-dust, anti-fly, rat proof, mothproof, washing, disposal of

waste water, and storage of garbage and waste;

3. Having professional food safety technicians and managerial personnel, and rules and regulations for ensuring the food safety;

4. Having reasonable equipment layout and technical flowchart so as to prevent cross pollution between the food to be processed and ready-to-eat food, and between raw materials and finished products, and to prevent the food from contacting with toxic substances or unclean articles;

5. Ensuring that the cutlery, drinking sets and containers for ready-to-eat food are washed clean or disinfected prior to use, the kitchenware and utensils are washed clean after use and kept clean;

6. Ensuring that the containers, utensils and equipment for storing, transporting, loading and unloading food are safe and innocuous, are kept clean so as to prevent pollution to food, reach the necessary temperature for food safety and meet other special requirements, and that the food may not be transported together with toxic or harmful articles;

7. Having small packages or using innocuous and clean packing materials or cutlery for the ready-to-eat food;

8. Ensuring that the persons engaging in the production or business operation of food shall keep personal hygiene, wash their hands clean and wear clean clothes and hats during the process of production or business operation, and that they use innocuous and clean vending devices when selling unpacked ready-to-eat food;

9. Using water which conforms to the national hygiene standards for drinking water;

10. Using detergent or disinfectant which is safe and innocuous to human body; and

11. Other requirements as prescribed in laws and regulations.

Article 28 It is forbidden to produce or engage in business operation of the following food:

1. Food produced with non-food raw materials, or food containing non-food-additive chemical substances and other substances potentially hazardous to human health, or food produced with recycled food as raw materi-

als;

2. Food in which the pathogenic microorganisms, pesticide residues, veterinary medicine residues, heavy metals, pollutants and other substances hazardous to human health exceed the limits as prescribed in the food safety standards;

3. Staple or supplementary food exclusively for infants and other particular groups of people, of which the nutrient ingredients do not meet the food safety standards;

4. Food that is putrid or deteriorated, spoiled by rancid oil or fat, moldy, infested with pest, contaminated and dirty, mixed with strange objects, adulterated and impure, or abnormal in sensory properties;

5. Meat of poultry, livestock, beasts and aquatic animals that died from disease or poisoning or for some unknown cause, and the products made from it;

6. Meat that has not been quarantined by the animal health inspection institution or has failed the quarantine or meat products that have not been inspected or have failed the inspection;

7. Food that is contaminated by packing materials, containers or transport vehicles;

8. Food whose shelf-life has expired;

9. Pre-packed food without labels;

10. Food, the production and business operation of which is expressly banned by the state for anti-disease purpose or for other special needs; and

11. Other food which does not conform to the food safety standards or requirements.

Article 29 The state shall adopt a licensing system for the food production and business operation. Those intending to engage in food production, food circulation or catering services shall obtain a license for food production, food circulation or catering services.

A food producer who has obtained a food production license is not required to obtain a food circulation license when selling self-produced food at its (his) production place. A catering service provider who has obtained a catering service license is not required to obtain food production and circulation licenses when selling self-made or self-processed food at its (his) catering service place. An individual farmer is not required to obtain the food circulation license when selling self-produced edible agricultural products.

To engage in food production or business operation, a small food production or processing workshop or a food vendor shall meet the food safety requirement of this Law, namely adapting to its production or business operation scale and conditions, so as to ensure that the food which it (he) produces or operates is hygienic, nontoxic and innocuous. The relevant departments shall intensify the supervision and administration of such small food production or processing workshops and food vendors. The specific measures shall be formulated in pursuance of this Law by the standing committee of the people's congress of the province, autonomous region or municipality directly under the Central Government.

Article 30 The people's government at or above the county level shall encourage small food production or processing workshops to improve their working conditions, and encourage food vendors to do business in such fixed establishments as centralized trade markets, and stores.

Article 31 The quality supervision departments, industry and commerce administrative departments and food and drug supervision and administration departments at and above the county level shall, pursuant to the Administrative License Law of the People's Republic of China, examine the relevant materials submitted by an applicant under the requirements in subparagraphs (1) through (4) of Article 27 of this Law, and where necessary, conduct an on-site inspection of the production and business operation place of the applicant. If the applicant meets the prescribed conditions, it shall decide to grant it (him) a license. If it (he) does not meet the prescribed conditions, it shall decide not to grant it (him) a license, and make an explanation in writing.

Article 32 An enterprise engaging in the production or business operation of food shall establish and improve its food safety management system, strengthen the training of its employees in respect to food safety knowledge, be provided with full-time or part-time food safety managers, do a good job in inspecting the food which it produces or operates, and carry out food production and business operation activities according to law.

Article 33 The state shall encourage enterprises engaging in production and business operation of food to meet the good manufacturing practice (GMP) and implement a hazard analysis and critical control point system (HACCP) so as to improve the food safety management level.

Where an enterprise engaging in production or business operation of food has passed the certification of good manufacturing practice (GMP), hazard analysis and critical control point system (HACCP), the certification agency shall conduct follow-up investigation according to law. If the enterprise no longer meets the certification requirements, it shall revoke the certification according to law and timely notify the quality supervision department, industry and commerce administrative department and food and drug supervision and administration department and make an announcement to the public. The certification agency shall not charge any fee for the follow-up investigation.

Article 34 A food producer or business operator shall establish and implement a handler health management system. No one who suffers from dysentery, typhoid, viral hepatitis or any other infectious disease of digestive tract, or active tuberculosis, or suppurative or exudative skin disease or any other disease that may affect the food safety shall engage in the work involving contact with ready-to-eat food.

A person engaging in the production or business operation of food shall be subject to a health examination every year, and shall not commence such work until he has obtained a health certificate.

Article 35 An edible agricultural produce producer shall, in accordance with the food safety standards and relevant provisions of the state, use pesticides, fertilizers, growth regulators, veterinary medicines, feeds, feed additives and other agricultural inputs. An enterprise or farmers' professional cooperative and economic organization engaging in the production of edible agricultural products shall establish a production record system for edible agricultural products.

The agriculture administrative department at or above the county level shall intensify the administration and guidance on the use of the agricultural inputs and establish a sound system for the safe use of agricultural inputs.

Article 36 When purchasing food raw materials, food additives and food-related products, a food producer shall check and verify the supplier's license and product compliance certification document. It shall, under the food safety standards, inspect the food raw materials, for which the supplier is unable to furnish a compliance certification document. It shall not purchase or use any food raw material, food additive or food-related product that does not conform to the food safety standards.

A food production enterprise shall establish a check and inspection record system for the purchased food raw materials, food additives and food-related products so as to faithfully record such contents as the names, specifications and quantities of the food raw materials, food additives and food-related products, names and contact information of the suppliers, and purchase dates.

The check and inspection records of the purchased food raw materials, food additives and food-related products shall be true and be preserved for at least 2 years.

Article 37 A food production enterprise shall establish a food ex-factory check record system so as to check the inspection certificates and the safety conditions of ex-factory food and faithfully record the name, specifications, quantity, production date, production batch number and inspection compliance certificate number of food, name and contact information of purchasers, date of sale, etc.

The food ex-factory check records shall be true and shall be kept for at least 2 years.

Article 38 A producer of food, food additives or food-related products shall, under the food safety standards, inspect the food, food additives or food-related products it produces, and shall not allow the ex-factory of or sell any food, food additive or food-related product unless it passes the inspection.

Article 39 When purchasing food, a food business operator shall check and verify the supplier's license and food compliance certification document.

An enterprise engaging in the business operation of food shall establish a check and inspection record system for the purchased food so as to faithfully record such contents as the name, specifications, quantity, production batch number, shelf-life of the food, name and contact information of the supplier, purchase date, etc.

The check and inspection records of the purchased food shall be true and be preserved for at least 2 years.

For an enterprise engaging in business operation of food by means of centralized distribution, the headquarters of the enterprise may, in a centralized manner, check and verify the suppliers' licenses and food compliance certification documents and make check records of the purchased food.

Article 40 A food business operator shall store food under the requirements for ensuring food safety, periodically check the food inventory and timely clear up the food which has gone bad or whose shelf life has expired.

Article 41 To store food in bulk, a food business operator shall give clear indications of the name, date of production, shelf life, name and contact information of the producer etc. of the food at the place of storage.

To sell food in bulk, a food business operator shall give clear indications of the name, date of production, shelf life of the food, name and contact information of the producer as well as the name and contact information of the business operator on the containers and external packages of the food in bulk.

Article 42 The packages of pre-packed food shall be labeled. A label shall indicate

1. The name, specifications, net content and production date;
2. A table of ingredients or components;
3. The name, address and contact information of the producer;
4. The shelf life;
5. The product standard code;

6. The storage requirements;

7. The common names in the national standards for the food additives used;

8. The serial number of the production license; and

9. Other matters required by laws, regulations or food safety standards.

The labels of staple and supplementary food exclusively for infants and other particular groups of people shall also bear indications of the main nutrient ingredients and contents thereof.

Article 43 The state shall adopt a licensing system for the production of food additives. The requirements and procedures for applying for a food additive production license shall be in conformity with the relevant provisions of the state on the administration of licenses for production of industrial products.

Article 44 Where an entity or individual intends to apply for engaging in food production by using new food raw materials or for engaging in the production of a new food additive or a new food-related product, it or he shall submit to the health administrative department of the State Council the safety assessment documents of the pertinent product. The health administrative department of the State Council shall, within 60 days from the date on which it receives the application, organize an examination of the safety assessment documents of the product. If the food safety requirements are satisfied, it shall decide to grant to the applicant a license and make an announcement. If the food safety requirements are not satisfied, it shall decide not to grant the applicant a license, and make an explanation in writing.

Article 45 No food additive may be listed in the scope of allowed use unless it is really technically necessary and has been proved as safe and reliable upon risk assessment. The health administrative department of the State Council shall, on the basis of the technical necessities and food safety risk assessment results, timely revise the standards for the varieties, extent of use and dosage of food additives.

Article 46 A food producer shall use food additives under food safety standards on the varieties, extent of use and dosages of food additives and shall not, during the process of food production, use any non-food-additive chemical substance or any other substance which is potentially hazardous to human health.

Article 47 Food additives shall have labels, instructions and packages. The labels and instructions shall indicate the matters as prescribed in subparagraphs 1 through 6, 8 and 9 of paragraph 1 of Article 42 of this Law, and the extent of use, dosage and use methods, and the labels shall bear an indication of the characters "FOOD ADDITIVE".

Article 48 No food or food additive labels or instructions shall contain any false or exaggerated content or involve such functions as disease prevention and treatment. The food producer shall assume legal liabilities for the representations of the labels or instructions.

The labels and instructions of food and food additives shall be clear and easily identifiable.

No food or food additive, which is not in conformity with the contents as indicated by its label or instructions, shall be placed on the market for sale.

Article 49 A food business operator shall sell pre-packed food according to the warning signs, warning instructions or notes for attention as given on the food labels or instructions.

Article 50 No medicine may be added to food under production or business operation, but substances that are traditionally both food and traditional Chinese medicinal materials may be added thereto. The list of substances that are traditionally both food and traditional Chinese medicinal materials shall be formulated and published by the health administrative department of the State Council.

Article 51 The state shall stringently supervise foods claimed to have particular effects on human health. The relevant supervision and administration departments shall perform their functions according to law and undertake the responsibilities. The concrete administrative measures shall be prescribed by the State Council.

No food claimed to have particular effects on human health shall cause any acute, sub-acute or chronic harm to the human health. The labels and instructions of such food shall not involve the effect of prevention or treatment of any disease, and the contents thereof shall be true and indicate applicable groups of people, inapplicable groups of people, effective ingredients or symbolic ingredients and contents thereof, etc. The effects and ingredients of a product shall be consistent with the indications in the labels and instructions.

Article 52 The sponsor of a centralized trade market, the lessor of counters or the organizer of a trade fair shall check the licenses of the food business operators admitted thereto, set down the food safety management responsibilities of the food business operators admitted thereto, regularly check their business operation environment and conditions. If it finds any food business operator who violates this Law, it shall timely stop the violation and promptly report it to the local industry and commerce administrative department or food and drug supervision and administration department at the county level.

If the sponsor of a centralized trade market, lessor of counters or organizer of a trade fair fails to perform the obligations as described in the preceding paragraph and any food safety accident occurs in the market, it shall bear several and joint liabilities.

Article 53 The state shall establish a food recall system. Where a food producer finds that any food it produces does not conform to the food safety standards, it shall promptly stop the production, recall all the food already placed on market for sale, notify the related producers, business operators and consumers and record the recall and notification information.

Where a food business operator finds that any food under its business operation does not conform to the food safety standards, it shall promptly stop the operations, notify the related producers, business operators and consumers, and record the stop of operation and notification information. If the food producer considers it necessary to recall the food, it shall recall it immediately.

The food producer shall make remedies to, make innocuous disposal of, destroy or take other measures against the recalled food, and report to the quality supervision department at or above the county level the information about the recall of food and about disposal of the recalled food.

If the food producer or business operator fails, under the provisions of this Article, to recall or stop the business operation of the food that does not meet the food safety standards, the quality supervision department, industry and commerce administrative department or food and drug supervision and administration department at or above the county level may order it to recall the food or stop the business operations.

Article 54 The contents of a food advertisement shall be true, shall not contain any falsehood or exaggeration, nor shall they involve the effect of prevention or treatment of any disease.

No food safety supervision and administration department or agency undertaking the function of food inspection, food industrial association or consumers' association may recommend any food to consumers by advertisements or by other means.

Article 55 Where a social group or any other organization or individual recommends food to consumers in a false advertisement, and thus impairs the legitimate rights and interests of consumers, it or he shall, along with the food producer or business operator, bear several and joint liabilities.

Article 56 The people's governments at all levels shall encourage mass production and chain business operation or distribution of food.

Chapter V Food Inspection

Article 57 A food inspection agency shall not engage in the food inspection activities until it has obtained the qualifications under relevant certification and accreditation provisions of the state, except it is otherwise provided for by law.

The qualification accreditation conditions and inspection requirements of food inspection agencies shall be prescribed by the health administrative department of the State Council.

A food inspection agency established upon approval of the relevant competent department of the State Council or has been accredited according to law prior to the implementation of this Law may carry on the food inspection activities under this Law.

Article 58 A food inspection shall be independently made by the inspector(s) designated by the food inspection agency.

The inspector(s) shall inspect the food under relevant laws, regulations, food safety standards and inspection requirements, respect science, scrupulously abide by professional ethics, ensure the objectiveness and impartiality of the issued inspection data and conclusion, and shall not issue any false inspection report.

Article 59 The food inspection shall be subject to the food inspection agency and inspector mutual accountability system. A food inspection report shall bear the official seal of the food inspection agency as well as the signature or seal of the inspector(s). The food inspection agency and inspector(s) shall be responsible for the issued food inspection report.

Article 60 No food safety supervision and administrative department shall exempt any food from inspection.

The quality supervision departments, industry and commerce administrative departments and food and drug supervision and administration departments at and above the county level shall regularly or irregularly make food inspections by taking samples. To make an inspection by taking samples, the randomly selected samples shall be purchased and no inspection fee or any other fee may be charged.

Where a quality supervision department, industry and commerce administrative department or food and drug supervision and administration department at or above the county level needs to inspect the food during its law enforcement work, it shall authorize a food inspection agency, which conforms to the provisions of this Law, to make the food inspection, and pay the relevant expenses. If it holds objections to the inspection conclusion, it may arrange a new inspection according to law.

Article 61 An enterprise engaging in the production or business operation of food may, by itself, inspect the food it produces, or authorize a food inspection agency, which conforms to the provisions of this Law, to do so.

Where a food industrial association or any other organization or a consumer needs to authorize a food inspection agency to make an inspection of food, it or he shall authorize a food inspection agency, which conforms to the provisions of this Law, to do so.

Chapter VI Import and Export of Food

Article 62 The imported food, food additives and food-related products shall conform to the national food safety standards of China.

The imported food shall be subject to the inspection of the entry/exit inspection and quarantine institution. If it passes the said inspection, the customs office shall release it upon the strength of the clearance certificate issued by the entry/exit inspection and quarantine institution.

Article 63 For the import of food which is not covered by the national food safety standards, or for the initial import of a new food additive or food-related product, the importer shall file an application with the health administrative department of the State Council and submit relevant safety assessment materials. The health administrative department of the State Council shall decide whether to grant the license in accordance with Article 44 of this Law and timely formulate corresponding national food safety standards.

Article 64 If a food safety accident occurring abroad may have an impact within China, or a serious food safety problem is found in any imported food, the entry/exit inspection and quarantine department of the state shall timely take the risk pre-warning measures or control measures, and notify the health administrative department, agricultural administrative department, industry and commerce administrative department and food and drug administrative department of the State Council. The departments that have received the notification shall timely take corresponding measures.

Article 65 An exporter or agent to export food to China shall go through the record-filing formalities at the entry/exit inspection and quarantine department of the state. An overseas food production enterprise to export food to China shall be registered at the entry/exit inspection and quarantine department of the state.

The entry/exit inspection and quarantine department of the state shall regularly announce the list of exports and agents who have made record filing, and the list of overseas food production enterprises registered.

Article 66 The imported pre-packed food shall have labels and instructions in Chinese. The labels and instructions shall conform to this Law, other relevant laws, administrative regulations and national food safety standards of China, and state the place of origin as well as the name, address and contact information of the domestic agent. No pre-packed food may be imported if it does not have labels and instructions in Chinese or if the labels and instructions do not conform to the provision of this Article.

Article 67 An importer shall establish a record system for the import and sale of food so as to faithfully record the name, specifications, quantity, date of production, production or import batch number, shelf life, name and contact information of the exporter and purchaser, date of delivery of the food, etc.

The records of import and sale of food shall be true and be preserved for at least 2 years.

Article 68 The food to be exported shall be subject to the supervision and sampling inspection of the entry/exit inspection and quarantine institution and shall be released by the customs office upon the strength of the clearance certificate issued by the entry/exit inspection and quarantine institution.

An export food production enterprise or planting and breeding plants of raw materials for the exported food shall go through the record-filing formalities at the entry/exit inspection and quarantine department of the state.

Article 69 The entry/exit inspection and quarantine department of the state shall collect and consolidate the safety information about the imported and exported food and timely notify the relevant departments, institutions and enterprises.

The entry/export inspection and quarantine department of the state shall establish records of the credit-standing of importers, exporters and export food production enterprises of the imported and exported food and publish them. It shall intensify the inspection and quarantine of the food imported and exported by the importers, exporters and export food production enterprises which have bad records.

Chapter VII Handling of Food Safety Accidents

Article 70 The State Council shall organize the formulation of a national food safety emergency response plan.

The people's government at or above the county level shall, under relevant laws and regulations, the food safety emergency response plan of the people's government at the higher level, and by taking into consideration the local actualities, work out a food safety emergency response plan for its own administrative region and submit it to the people's government at the next higher level for archival purposes.

An enterprise engaging in production and business operation of food shall work out a plan on handling food safety accidents and regularly check the implementation of its own food safety prevention measures so as to timely eradicate the potential risks of food safety accident.

Article 71 The entity, in which a food safety accident occurs, shall deal with the accident immediately so as to prevent it from becoming worse. The entity, in which an accident occurs, and the entities receiving patients for medical treatment shall timely report the relevant situation to the health administrative department at the county level at the place of accident.

If the agriculture administrative department, quality supervision department, industry and commerce administrative department or food and drug supervision and administration department finds, during its routine supervision and administration, any food safety accident or receives any tip-off of food safety accident, it shall notify the health administrative department immediately.

At the occurrence of a major food safety accident, the health administrative department at the county level which receives the report shall, under relevant provisions, report to the people's government at the same level and to the health administrative department of the people's government at the higher level. The people's government at the county level and the health administrative department of the people's government at the higher level shall report it to its superior under relevant provisions.

No entity or individual may conceal, make false report or delay the report of any food safety accident, or destroy relevant evidence.

Article 72 As soon as a health administrative department at or above the county level receives a report of food safety accident, it shall, jointly with the agriculture administrative department, quality supervision department, industry and commerce administrative department and food and drug supervision and administration department, investigate and deal with it, and take the following measures to prevent or mitigate its hazards to the society:

1. To carry out the emergency response and rescue work; the health administrative department shall immediately organize the rescue and medical treatment of persons suffering personal injuries in a food safety accident;

2. To seal up the food and its raw materials which may result in the food safety accident and make an inspection immediately; to order the food producer or business operator to, under Article 53 of this Law, recall, stop business operation of and destroy the contaminated food and raw materials upon confirmation;

3. To seal up the utensils and devices used for the contaminated food, and order to have them cleaned and disinfected; and

4. To do a good job in releasing information, releasing the information about the food safety accident and about the handling of the accident according to law, and making explanations and statements about the possible hazards.

In the case of a major food safety accident, the people's government at or above the county level shall promptly form a command body for handling the food safety accident, initiate the emergency plan and deal with the accident in accordance with the provision of the preceding paragraph.

Article 73 In the case of a major food safety accident, the health administrative department of the people's government at or above the level of a districted city shall, jointly with the relevant departments, investigate the liabilities for the accident, urge the relevant departments to perform their functions, and report to the people's government at the same level a report about the investigation and handling of accident liabilities.

Where a major food safety accident involves 2 or more provinces, autonomous regions and municipalities directly under the Central Government, the health administrative department of the State Council shall organize an investigation of the accident liabilities according the provisions of the preceding paragraph.

Article 74 At the occurrence of a food safety accident, the disease prevention and control institution shall assist the health administrative department and other relevant departments in performing the sanitary treatment at the scene of accident and conduct an epidemiological investigation into the factors relating to the food safety accident.

Article 75 In the investigation of a food safety accident, in addition to the liabilities of the entity in which the accident occurs, the neglect or dereliction of duty on the side of functionaries of the supervision and administration department or of the certification agency having the functions of supervision and administration or certification shall be found out.

Chapter VIII Supervision and Administration

Article 76 A local people's government at or above the county level shall organize the health administrative department, agriculture administrative department, quality supervision department, industry and commerce administrative department and food and drug supervision and administration department at the same level to work out an annual plan on the food safety supervision and administration of its own administrative region and carry out the relevant work under the annual plan.

Article 77 The quality supervision department, industry and commerce administrative department and food and drug supervision and administration department at or above the county level shall perform their respective functions on food safety supervision and administration, and have the power to take the following measures:

1. To conduct on-site inspections by entering the production and business operation sites;

2. To conduct sampling inspection on the food under production and business operation;

3. To consult and copy relevant contracts, instruments, account books and other relevant materials;

4. To seal up and detain the food that, as evidence shows, does not conform to the food safety standards, the food raw materials, food additives and food-related products for illegal use, as well as the utensils and equipment that are used for illegal production and business operation or that have been contaminated; and

5. To seal up the sites for the illegal production and business operation of food;

The agriculture administrative department at or above the county level shall, in accordance with the Law of the People's Republic of China on Agricultural Product Quality Safety, supervise and administer the edible agricultural products.

Article 78 When supervising and inspecting the food producers and business operators, a quality supervision department, industry and commerce administrative department or food and drug supervision and administration department at or above the county level shall record the supervision and inspection information as well as the handling results. The supervision and inspection records shall be archived after being signed by the supervision and inspection personnel as well as by the food producer or business operator.

Article 79 A quality supervision department, industry and commerce administrative department or food and drug supervision and administration department at or above the county level shall establish food safety credit archives for food producers and business operators so as to record the information about the issuance of licenses, routine supervision and inspection results, investigation and handling of unlawful conducts, etc., and shall, in light of the records in the food safety credit archives, increase the frequency of supervision and inspection on food producers and business operators having bad credit records.

Article 80 Where a health administrative department, quality supervision department, industry and commerce administrative department or food and drug supervision and administration department at or above the county level receives a consultation request, complaint or tip-off, it shall accept it if it falls within the scope of its functions, and shall timely make a reply, verify and deal with it. If it does not fall within the scope of its functions, it shall give the party concerned a written notice and transfer the case to the competent department. The competent department shall timely deal with it, and shall not decline it. If it is a food safety accident, it shall be handled under the relevant provisions of Chapter VII of this Law.

Article 81 A health administrative department, quality supervision department, industry and commerce administrative department or food and drug supervision and administration department at or above the county level shall, under the statutory functions and procedures, perform the food safety supervision and administration functions. It shall not impose the administrative punishment of fine twice or more against the same unlawful conduct of a producer or business operator. If the producer or business operator is suspected of committing any crime, it shall transfer the case to the public security organization according to law.

Article 82 The State shall establish a uniform system for the release of food safety information. The following information shall be uniformly released by the health administrative department of the State Council:

1. The overall information about the national food safety;

2. The food safety risk assessment information and food safety risk warning information;

3. The information about major food safety accidents and about the han-

dling thereof; and

4. Other important food safety information, and the information which the State Council determines necessary to release uniformly.

For the information as described in sub-paragraphs 2 and 3, if its consequences are limited to a specific region, it may be released by the health administrative department of the people's government of the relevant province, autonomous region or municipality directly under the Central Government. A health administrative department, quality supervision department, industry and commerce administrative department or food and drug supervision and administration department at or above the county level shall, in light of its own functions, release the information about its routine supervision and administration of food safety.

A food safety supervision and administration department shall ensure the accuracy, timeliness and objectiveness of the information it releases.

Article 83 When a local health administrative department, agriculture administrative department, quality supervision department, industry and commerce administrative department or food and drug supervision and administration department at or above the county level gets any information which is required to be uniformly released according to paragraph 1 of Article 82 of this Law, it shall promptly report to its superior administrative department. Its superior administrative department shall promptly report to the health administrative department of the State Council. Or it even may, where necessary, directly report to the health administrative department of the State Council.

The health administrative department, agriculture administrative department, quality supervision department, industry and commerce administrative department and food and drug supervision and administration department at or above the county level shall notify each other of the food safety information they get.

Chapter IX Legal Liabilities

Article 84 Where a violator of this Law engages in the food production or business operation activities without a license or produces food additives without a license, its illegal gains, food or food additives under its illegal production and business operation, as well as the utensils, equipment, raw materials and other articles used for the illegal production or business operation shall be confiscated by the relevant competent departments under their respective functions. If the monetary value of the illegally produced or operated food or food additives is less than 10,000 yuan, the violator shall be fined not less than 2,000 yuan but not more than 50,000 yuan concurrently. If the monetary value of the said food or food additive is 10,000 yuan or more, the violator shall be fined not less than 5 times but not more then 10 times the monetary value concurrently.

Article 85 Where a violator of this Law is under any of the following circumstances, its illegal gains, illegally produced or operated food, and utensils, equipment, raw materials and other articles used for the illegal production or business operation shall be confiscated by the relevant competent departments under their respective functions. If the monetary value of the illegally produced or operated food is less than 10,000 yuan, the violator shall be fined not less than 2,000 yuan but not more than 50,000 yuan concurrently. If the monetary value of the said food is 10,000 yuan or more, the violator shall be fined not less than 5 times but not more then 10 times the monetary value concurrently. If the circumstance is serious, the license of the violator shall be revoked:

1. It produces food with non-food raw materials, or food containing non-food-additive chemical substances and other substances potentially hazardous to human health, or food produced with recycled food as raw materials;

2. It produces or engages in the business operation of food in which the pathogenic microorganisms, pesticide residues, veterinary medicine residues, heavy metals, pollutants and other substances hazardous to human

health exceed the limits as prescribed in the food safety standards;

3. It produces or engages in the business operation of staple or supplementary food exclusively for infants or other particular groups of people, of which the nutrient ingredients do not meet the food safety standards;

4. It engages in the business operation of food that is putrid or deteriorated, spoiled by rancid oil or fat, moldy, infested with pest, contaminated and dirty, mixed with strange objects, adulterated and impure, or abnormal in sensory properties;

5. It engages in the business operation of the meat of poultry, livestock, beasts and aquatic animals that died from disease or poisoning or for some unknown cause, or such meat products;

6. It engages in the business operation of the meat that has not been quarantined or has failed the quarantine by the animal health inspection institution or meat products that have not been inspected or have failed the inspection;

7. It engages in the business operation of the food of which the shelf life has expired;

8. It produces or engages in the business operation of food, the production and business operation of which is expressly banned by the state for anti-disease purpose or for other special reasons;

9. It produces food with new food raw materials or produces a new food additive or new food-related product without undergoing the safety assessment; or

10. The food producer or business operator still refuses to recall or stop the business operation of the food which does not conform to the food safety standards, after the relevant competent department so orders.

Article 86 Where a violator of this Law is under any of the following circumstances, its illegal gains, illegally produced or operated food, utensils, equipment, raw materials and other articles used for the illegal production or business operation shall be confiscated by the relevant competent departments under their respective functions. If the monetary value of the illegally produced or operated food is less than 10,000 yuan, the violator shall be fined not less than 2,000 yuan but not more than 50,000 yuan concurrently. If the monetary value of the said food is 10,000 yuan or more, the violator shall be fined not less than 2 times but not more then 5 times the monetary value concurrently. If the circumstance is serious, the violator shall be ordered to stop production or business operation or even have its business license revoked:

1. It engages in the business operation of the food contaminated by packing materials, containers, transport means, etc.;

2. It produces or engages in the business operation of pre-packed food or food additive without labels, or produces or engages the business operation of food or food additive of which the labels or instructions do not conform to the provisions of this Law;

3. The food producer purchases and uses food raw materials, food additives or food-related products which do not conform to the food safety standards; or

4. The food producer or business operator adds any medicine to the food.

Article 87 Where a violator of this Law is under any of the following circumstances, the relevant competent departments shall, under their respective functions, order it to make a correction and give it a warning. If it refuses to make a correction, it shall be fined not less than 2,000 yuan but not more than 20,000 yuan. If the circumstance is serious, it shall be ordered to stop production and business operation, or its license shall be revoked even:

1. Failing to inspect the food raw materials purchased by it and the food, food additives and food-related products produced by it;

2. Failing to establish and abide by the check and inspection record system or ex-factory check record system;

3. Having laid down enterprise food safety standards but failing to go through the record-filing formalities under this Law;

4. Failing to store or sell food or clear up the food inventory under the prescribed requirements;

5. Failing to check the license and relevant certification documents when purchasing goods;

6. The labels or instructions of the food or food additive produced involve the effect of prevention or treatment of any disease; or

7. Assigning any person, who suffers from any of the diseases as listed in Article 34 of this Law, to engage in the work involving contact with ready-to-eat food.

Article 88 After the occurrence of a food safety accident, if the entity in which the food safety accident occurs fails, by violating this Law, to handle or report the accident, it shall be ordered to make a correction and be given a warning by the relevant competent departments under their respective functions. If it destroys relevant evidence, it shall be ordered to stop production or business operation and concurrently be fined not less than 2,000 yuan but not more than 100,000 yuan. If it causes any severe consequences, its license shall be revoked by the original issuing department.

Article 89 Where a violator of this Law is under any of the following circumstances, it shall be punished under Article 85 of this Law:

1. It imports any food which does not conform to the national food safety standards of China;

2. It imports any food which is not covered by the national food safety standards, or initially imports any new food additive or food-related product without undergoing the safety assessment; or

3. The exporter exports food by violating this Law.

An importer, which violates this Law due to its failure to establish or failure to observe the food import and sale record system, shall be punished in accordance with Article 87 of this Law.

Article 90 Where, in violation of this Law, the sponsor of a centralized trade market, the lessor of counters or the organizer of a trade fair allows a food business operator without a license to enter the market to sell food, or fails to perform such obligations as inspection and reporting, it shall be fined not less than 2,000 yuan but not more than 50,000 yuan by the competent departments under their respective functions. If severe consequences are caused, it shall be ordered to stop business operation, and have its license revoked by the original issuing department.

Article 91 A violator of this Law who fails to follow the relevant requirements in the transport of food shall be ordered to make a correction and be given a warning by the relevant competent departments under their respective functions. If it refuses to make a correction, it shall be ordered to stop production or business operation, and be fined not less than 2,000 yuan but not more than 50,000 yuan. If the circumstance is serious, it shall have its license revoked by the original issuing department.

Article 92 The directly responsible person-in-charge of an entity whose license for food production, circulation or catering services is revoked shall not engage in the management of food production and business operation within 5 years as of the date on which the punishment decision is made.

Where a food producer or business operator hires a person, who is forbidden to engage in the management of food production and business operation, to engage in the management work, its license shall be revoked by the original issuing department.

Article 93 Where a food inspection agency or food inspector violates this Law due to issuing any false inspection report, the competent department or institution, which granted it the qualification, shall revoke its inspection qualification and shall, according to law, give the directly responsible person-in-charge and the food inspector a sanction of removal or dismissal.

A person who is subject to a criminal punishment or sanction of dismissal because of his violation of this Law shall not engage in the food inspection work within 10 years as of the date on which the execution of the criminal punishment is ended or the sanction decision is made. Where a food inspection agency hires any person forbidden to engage in the food inspection work, the competent department or institution, which granted it the qualification, shall revoke its inspection qualification.

Article 94 A violator of this Law who makes misrepresentations about the food quality in an advertisement and thus misleads the consumers shall be punished in accordance with the Advertising Law of the People's Republic of China.

Where a food safety supervision and administration department or agency undertaking the function of food inspection, food industrial association or consumers' association recommends, in violation of this Law, any food to consumers by advertisements or by other means, the relevant competent department shall, in pursuance of law, confiscate its illegal gains, impose such sanctions as major demerit, demotion or removal on the directly responsible person-in-charge and other directly liable persons.

Article 95 Where a local people's government at or above the county level fails, in violation of this Law, to perform its functions in the food safety supervision and administration, and any major food safety accident has occurred within its administrative region and causes severe social consequences, the directly responsible person-in-charge and other directly liable persons shall, according to law, be given a sanction of major demerit, demotion or removal or dismissal.

Where a health administrative department, agriculture administrative department, quality supervision department, industry and commerce administrative department, food and drug supervision and administration department or any other relevant administrative department at or above the county level fails, in violation of this Law, to perform the functions as prescribed in this Law or abuses its powers, neglects its duties or practices favoritism, the directly responsible person-in-charge and other directly liable persons shall be given a sanction of major merit or demotion according to law. If severe consequences are caused, they shall be given a sanction of removal or dismissal and the major person-in-charge shall take the blame and resign.

Article 96 A violator of this Law who causes personal, property or other damages shall bear the compensation liability.

Besides claiming damages, a consumer may require the producer, who produces food which does not conform to the food safety standards, or the seller who knowingly sells food which does not conform to the food safety standards, to pay 10 times the money paid.

Article 97 A violator of this Law shall bear the civil compensation liability and pay the fine or pecuniary penalty. If its (his) property is insufficient to cover all the payment at the same time, it (he) shall first bear the civil compensation liability.

Article 98 For a violator of this Law, if any crime is constituted, it (he) shall be subject to the criminal liabilities.

Chapter X Supplementary Provisions

Article 99 Definitions of the following terms as used in this Law:

The term "food" refers to the finished products and raw materials for people to eat or drink, and articles which are traditionally food and medicine, excluding articles that are used for the purpose of medical treatment.

The term "food safety" means that the food is nontoxic, innocuous and satisfies the necessary nutritional requirements, and does not cause any acute, sub-acute or chronic hazards to the human health.

The term "pre-packed food" refers to the food of fixed quantity which is packed or made in packing materials and containers in advance.

The term "food additive" refers to any synthetic or natural substance that is added to food for improving its quality, color, flavor or taste, or for the needs of inhibiting spoilage, preservation or processing.

The term "packing materials and containers of food" refers to the paper, bamboo, wood, metal, enamel, ceramic, plastic, rubber, natural fiber, chemical fiber, glass and other products used for packing and containing food or food additives, and the paints that directly contact food or food additives.

The term "utensils and equipment for food production or business operation" refers to the machinery, pipes, conveyors, containers, utensils, cutlery, etc. that directly contact food or food additives during the course of production, circulation and utilization of food or food additives.

The term "detergent or disinfectant used for food" refers to the substances that are directly used for washing or disinfecting food, cutlery, drinking sets, and utensils, equipment or food packing materials and containers directly contacting the food.

The term "shelf life" refers to the term of quality guarantee of pre-packed food under the storage conditions as stated on its labels.

The term "food-borne disease" refers to an infectious or poisoning disease or any other disease resulting from the entry of pathogenic factors of food into the human body.

The term "food poisoning" refers to the acute or sub-acute disease occurring after the eating of food contaminated by toxic or harmful substances or food containing toxic and harmful substances.

The term "food safety accident" refers to an accident that stems from food and is or may be hazardous to the human body, such as food poisoning, food-borne disease or food contamination.

Article 100 The corresponding license which a food producer or business operator has already obtained prior to the implementation of this Law shall remain valid.

Article 101 The food safety administration of dairy products, genetically modified food, slaughtering of live pigs, spirits and common salt shall be governed by this Law. If it is otherwise provided for by any other law or administrative regulation, that law or administrative regulation shall prevail.

Article 102 The administrative measures for the food safety in the railway business operations shall be formulated by the health administrative department of the State Council jointly with the relevant department of the State Council in pursuance of this Law.

The administrative measures for the food safety of the exclusive food and self-supplied food of the army shall be formulated by the Central Military Commission according to this Law.

Article 103 Where necessary, the State Council may make adjustments to the food safety supervision and administration system.

Article 104 This Law shall come into force as of June 1, 2009. The Food Hygiene Law of the People's Republic of China shall be abolished simultaneously.

Law of the People's Republic of China on Agricultural Product Quality Safety

(Adopted by the 21st Meeting of the Standing Committee of the 10th National People's Congress on April 29, 2006 and effective as of November 1, 2006)

Chapter I General Provisions

Article 1 The present Law is formulated in order to guarantee the quality safety of agricultural products, maintain the health of the general public and promote the development of agriculture and rural economy.

Article 2 The term "agricultural products" as mentioned in the present Law refers to primary products sourced from agriculture, that is to say, the plants, animals, microorganisms and their products which are obtained from agricultural activities.

The term "agricultural product quality safety" as mentioned in the present Law means that the quality of an agricultural product meets the requirements of ensuring human health and safety.

Article 3 The administrative department of agriculture of the people's government at the county level or above shall be responsible for the supervision and inspection of agricultural product quality safety; while the relevant departments of the people's government at the county level or above shall, in accordance with the scope of duties, be responsible for the relevant work on agricultural product quality safety respectively.

Article 4 The people's government at the county level or above shall include agricultural product quality safety administration into the national economic and social development planning at the present level and offer funds of agricultural product quality safety for carrying out the work of agricultural product quality safety.

Article 5 The local people's government at the county level or above shall exercise the unified leadership over and coordinate the work of agricultural product quality safety under their own jurisdiction, adopt measures to set up and perfect an agricultural product quality safety service system and improve the level of agricultural product quality safety.

Article 6 The administrative department of agriculture of the State Council shall set up an agricultural product quality safety risk evaluation experts committee consisting of experts in relevant areas, in order to carry out risk analysis and evaluation of the potential harms which might affect the agricultural product quality safety.

The administrative department of agriculture of the State Council shall adopt relevant administrative measures in accordance with the risk evaluation results of agricultural product quality safety and inform the relevant departments of the State Council of the risk evaluation results of agricultural product quality safety in a timely manner.

Article 7 The administrative department of agriculture of the State Council and the administrative department of agriculture of the people's government of each province, autonomous region or municipality directly under the Central Government shall publish relative information on the situation of agricultural product quality safety according to their legal authority.

Article 8 The state guides and popularizes standardized production of agricultural products, encourages and supports production of high-quality agricultural products, and prohibits production and sales of agricultural products which do not meet the agricultural product quality safety criteria prescribed by the state.

Article 9 The state supports scientific and technological researches on agricultural product quality safety, implements scientific quality safety administration methods and promotes advanced and safe production technologies.

Article 10 The people's government at any level and the relevant departments shall strengthen publicity of knowledge on agricultural product quality safety, improve the consciousness on agricultural product quality safety of the general public, guide producers and sellers of agricultural products to intensify quality safety management and guarantee the safety of agricultural product consumption.

Chapter II Agricultural Product Quality Safety Criteria

Article 11 The state establishes and improves the system of agricultural product quality safety criteria. The agricultural product quality safety criteria shall be compulsory technical norms.

The agricultural product quality safety criteria shall be formulated and promulgated in light of relative laws and administrative regulations.

Article 12 When formulating the agricultural product quality safety criteria, the relevant departments shall take into full consideration the risk evaluation results of agricultural product quality safety and give audience to the opinions of producers, sellers and consumers of agricultural products, in order to guarantee the consumption safety.

Article 13 The agricultural product quality safety criteria shall be revised in a timely manner based on the scientific and technological development level and the requirements of agricultural product quality safety.

Article 14 The agricultural product quality safety criteria shall be organized to implement by the administrative department of agriculture together with other relevant departments.

Chapter III Producing Areas of Agricultural Products

Article 15 The administrative department of agriculture of a local people's government at the county level or above shall, as per the requirements of agricultural product quality safety as well as in accordance with factors such as variety characters of the agricultural products and poisonous and harmful substances in the atmosphere, soil and water body of the production area, propose areas banned from production which it considers unsuitable for production of certain agricultural products, and publicize such areas upon approval of the people's government at the same level. The specific measures shall be formulated by the administrative department of agriculture of the State Council together with the administrative department of environmental protection of the State Council.

The adjustment of areas banned from production of agricultural products shall be made in light of the procedures prescribed in the preceding paragraph.

Article 16 The people's government at the county level or above shall adopt measures to enhance construction of agricultural product bases and improve the production conditions of agricultural products.

Measures shall be taken by the administrative department of agriculture of the people's government at the county level or above to propel construction of comprehensive demonstration areas for standardized production, demonstration farms, breeding areas and areas without prescribed epidemic animal or plant diseases, so as to guarantee the agricultural products quality safety.

Article 17 It is prohibited to produce, fish or collect edible agricultural products or to establish production bases of agricultural products in the areas where poisonous and harmful substances are in excess of the prescribed standards.

Article 18 It is prohibited to discharge or dump waste water, waste gas, solid wastes or other poisonous and harmful substances to producing areas of agricultural products in violation of laws and regulations.

The water used for agricultural production and the solid wastes used as fertilizers shall meet the criteria of the state provisions.

Article 19 Such chemical products as chemical fertilizers, pesticides, veterinary drugs and agricultural films shall be used in a reasonable way by producers of agricultural products to prevent such chemical products from polluting the producing areas of agricultural products.

Chapter IV Production of Agricultural Products

Article 20 The requirements on production technologies and operational rules shall be constituted by the administrative department of agriculture of the State Council and the administrative department of agriculture of the people's government of each province, autonomous region or municipality directly under the Central Government so as to guarantee the agricultural product quality safety. The administrative department of agriculture of each people's government at the county level or above shall strengthen its guidance to the production of agricultural products.

Article 21 For the pesticides, veterinary drugs, feeds and feed additives, fertilizers and veterinary devices, which might affect agricultural product quality safety, a licensing system shall be carried out in light of relative laws and administrative regulations.

The administrative department of agriculture of the State Council and the administrative department of agriculture of the people's government of each province, autonomous region or municipality directly under the Central Government shall, at a regular time schedule, make a random inspection on such agricultural input products as pesticides, veterinary drugs, feeds and feed additives as well as fertilizers, which might endanger the agricultural product quality safety, and shall make public the results.

Article 22 The administrative department of agriculture of the people's government at the county level or above shall enhance administration and guidance on the use of agricultural input products, as well as setting up and improving a system for safe use of agricultural input products.

Article 23 Agricultural research and education institutions and agricultural technology promotion institutions shall strengthen trainings on quality safety knowledge and skills for producers of agricultural products.

Article 24 An enterprise engaging in agricultural production or a professional farmers cooperative economic organization shall set up records on production of agricultural products and the contents as follows shall be included:

(1) The names, sources, usage, dosage of agricultural input products in use, the date of using it and the date disusing it;

(2) The information on occurrence, prevention and control of animal epidemic diseases as well as plant diseases, pests and disasters; and

(3) The date of harvest, slaughter or fishing.

The records on agricultural production shall be preserved for two years. Any forgery of records on agricultural production is prohibited. The state encourages other producers engaging in agricultural production to set up records on agricultural production.

Article 25 A producer engaging in agricultural production shall, in light of the laws, administrative regulations and provisions of the administrative department of agriculture of the State Council, make use of the agricultural input products in a reasonable way, strictly carring out the provisions on safe intervals or withdrawal period for using agricultural input products, so as to prevent the agricultural input products from endangering the agricultural product quality safety.

Any agricultural input product prohibited by explicit order of the state shall be forbidden to be used in the process of agricultural production.

Article 26 An enterprise engaging in agricultural production or a professional farmers cooperative economic organization shall check the agricultural product quality safety either by itself or by entrusting a testing institution. It is prohibited to sell any agricultural product found from the test to fail to comply with the agricultural product quality safety criteria .

Article 27 A professional farmers cooperative economic organization or an agricultural products industry association shall offer its members production technology services in a timely manner, set up agricultural product quality safety management systems, perfect the agricultural product quality safety control system and strengthen self-disciplinary management.

Chapter V Packages and Marks of Agricultural Products

Article 28 Where the agricultural products sold by an enterprise engaging in production of agricultural products, a professional farmers cooperative economic organization or an entity or an individual engaging in purchase of agricultural products are required in accordance with relevant provisions to be packed or be attached with marks, they may not be sold until they have been packed or attached with marks. Such contents as the product name, producing area, producer, date of production, warranty period and product quality grade shall be indicated on the packages or marks, in accordance with related provisions; if any additive is used, the name of the additive shall also be indicated in accordance with the provisions. The specific measures shall be instituted by the administrative department of agriculture of the State Council.

Article 29 The materials used in package, preservation, storage and transport of agricultural products, such as preservatives, antiseptics, additives, etc., shall comply with the relevant compulsory technical norms of the state.

Article 30 The agricultural products belonging to agricultural transgenic organisms shall be marked in light of relative provisions on the administration of the safety of agricultural transgenic organisms.

Article 31 The animals, plants and their products required to be quarantined in accordance with the law shall be attached with quarantine marks of fitness and quarantine certificates of fitness.

Article 32 The on-sale agricultural products must meet the agricultural product quality safety criteria, and the producers may submit applications for using pollution-free marks on agricultural products. If the quality of the agricultural products complies with the criteria prescribed by the state for high-quality agricultural products, the producers may submit applications for using commensurate quality marks on agricultural products.

It is prohibited to imitate the quality marks on agricultural products as prescribed in the preceding paragraph.

Chapter VI Supervision and Inspection

Article 33 An agricultural product under any of the following circumstances shall not be sold:

(1) It contains any pesticide, veterinary drug or other chemical substance prohibited by the state from being used;

(2) The remnant of chemical substance such as pesticide and veterinary drug or the contained poisonous and harmful substance such as heavy metal, etc. does not comply with the agricultural product quality safety criteria;

(3) The contained pathogenic parasites, microorganisms or biological toxin do not conform to the agricultural product quality safety criteria;

(4) The material in use such as preservative, antiseptic or additive, etc. does not conform to the relative compulsory technical norms of the state; or

(5) Other circumstances under which it does not conform to the agricultural product quality safety criteria.

Article 34 The state sets up an agricultural product quality safety monitoring system. The administrative department of agriculture of the people's government at the county level or above shall, in accordance with the requirements for guaranteeing the agricultural product quality safety, make a plan of monitoring the agricultural product quality safety, organize the implementation thereof and supervise and make a random inspection on the agricultural products under production or on sale in the market. The administrative department of agriculture of the State Council or the administrative department of agriculture of the people's government of each province, autonomous region or municipality directly under the Central Government shall make public the results according to its legal authority.

For a supervisory test on a random inspection, the department concerned shall entrust an agricultural product quality safety test institution that meets the conditions in Article 35 of the present Law, but shall not charge any fee from the party to be tested. The number of the samples shall not exceed the quantity prescribed by the administrative department of agriculture of the State Council. For the agricultural products which are subject to supervision of the administrative departments of agriculture at higher levels by sampling, the administrative departments of agriculture at lower levels shall not make a sampling again.

Article 35 For the agricultural product quality safety test, the existing qualified test institutions shall be given full consideration.

An institution engaging in agricultural product quality safety test must possess commensurate conditions and capacities for test and shall be qualified and pass the assessment of the administrative department of agriculture of the people's government at the provincial level or above or its authorized department. The detailed measures shall be instituted by the administrative department of agriculture of the State Council.

An agricultural product quality safety test institution shall be found qualified from metrological certification in accordance with law.

Article 36 Where a producer or seller of agricultural products has any objection to the result of random inspection, it may, within five days as of the receipt of the test result, submit an application to the administrative department of agriculture that organizes the implementation of the random inspection on agricultural product quality safety or to the administrative departments of agriculture at higher levels for a re-test.

When the speedy test method ascertained by the administrative department of agriculture of the State Council together with the relevant departments is adopted for the random inspection on agricultural product quality safety, if the party that is tested has any objection to the test result, it may, within four hours as of the receipt of the test result, submit an application for a re-test. The re-test shall not be carried out in a speedy method.

If the testing institution causes any damages to the party concerned because of a wrong test result, it shall undertake liabilities for compensation in light of the law.

Article 37 An agricultural product wholesale market shall establish or entrust an agricultural product quality safety test institution to test the quality safety of the agricultural products sold in the market by random inspection; when finding any inconformity with the agricultural product quality safety criteria, it shall require the seller to immediately stop the sale and report to the administrative department of agriculture.

An enterprise engaging in sale of agricultural products shall, for the agricultural products it sells, set up and improve the rules on inspection and acceptance of purchased goods; any agricultural product that is found from the inspection to fail to comply with the quality safety criteria shall not be sold.

Article 38 The state encourages entities and individuals to carry out public supervision over the agricultural product quality safety. Any entity or individual shall have the right to impeach, expose or accuse any act violating the present Law. After receipt of relevant impeachment, exposure or accusation, the relevant department shall deal with the case in a timely manner.

Article 39 The administrative department of agriculture of the people's government at the county level or above may, in its agricultural product quality safety supervision and inspection, make on-site inspections on the agricultural products under production or on sale, investigate and know about the relative information on agricultural product quality safety, consult and photocopy the records and other information concerning agricultural product quality safety; and shall have the right to seal up or distrain the agricultural products which are found from test to fail to comply with the agricultural product quality safety criteria.

Article 40 When an agricultural product quality safety accident occurs, the concerned entities and individuals shall take control measures and report to the local people's government at the township level and the administrative department of agriculture of the people's government at the county level in a timely manner. The organ receiving the report shall deal with the accident in a timely manner and report to the people's government at the higher level and other relative departments. When a significant agricultural product quality safety accident occurs, the administrative department of agriculture shall inform the food and drug administrative department at the same level of the accident in a timely manner.

Article 41 In the agricultural product quality safety supervision and administration, if the administrative department of agriculture of a people's government at the county level or above finds that an agricultural product is under any of the circumstances listed in Article 33 of the present Law, it shall, on the basis of the requirements of the system for ascertaining liabilities of agricultural product quality safety, find out the liable persons and decide punishment in light of the law or propose punishment suggestions.

Article 42 An imported agricultural product must be inspected in accordance with the agricultural product quality safety criteria prescribed by the state. If the relevant agricultural product quality safety criteria have not been formulated, the department concerned shall formulate them in light of the law in a timely manner and may, before finishing formulating such criteria, inspect the imported agricultural product by referring to the relevant foreign criteria designated by the relevant department of the state.

Article 43 If any agricultural product quality safety supervisory and administrative staff member does not carry out his supervisory duties in light of the law or abuses his power, he shall be given administrative sanctions according to law.

Chapter VII Legal Liabilities

Article 44 If an agricultural product quality safety test institution forges a test result, it shall be ordered to make correction. Its illegal proceeds shall be confiscated and in addition, it shall be charged a fine not less than 50,000 yuan and not more than 100,000 yuan. The person-in-charge directly responsible and other persons held direct liabilities shall be charged a fine not less than 10,000 yuan and not more than 50,000 yuan respectively. If the circumstances are serious, its test qualification shall be revoked. If it causes any damages, it shall undertake liabilities for compensation in light of the law.

If an agricultural product quality safety test institution issues an untrue test result and causes any damages, it shall undertake liabilities for compensation in light of the law; if it causes any heavy damages, its test qualification shall be revoked in addition.

Article 45 Whoever violates laws or regulations by discharging or dumping waste water, waste gas, solid wastes or other poisonous and harmful substances to a producing area of agricultural products shall be penalized in light of the relative environmental protection laws and regulations; if he causes any damage, he shall undertake liabilities for compensation in light of the law.

Article 46 Whoever violates laws, administrative regulations or any provisions of the administrative department of agriculture of the State Council in using agricultural input products shall be penalized in light of the relative laws and administrative regulations.

Article 47 If an enterprise engaging in agricultural production or a professional farmers cooperative economic organization fails to establish or preserve records on agricultural production according to related provisions, or forges records on agricultural production, it shall be ordered to make correction within a time limit; if it fails to make correction within the time limit, it may be fined not more than 2,000 yuan.

Article 48 Whoever violates the provisions prescribed in Article 28 of the present Law by failing to follow the provisions to pack or mark the agricultural products for sale shall be ordered to make correction within a time limit; if he fails to make correction within the time limit, he may be fined not more than 2,000 yuan.

Article 49 If any of the circumstances under Item (4) of Article 33 of the present Law arises and the material in use such as the preservative, antiseptic or additive, etc. does not comply with the relevant compulsory technical norms of the state, the party concerned shall be ordered to stop selling the agricultural products and shall make innocuous treatment of the polluted agricultural products. If no innocuous treatment can be made, the agricultural products shall be destroyed under supervision; at the same time, his illegal income shall be confiscated and he shall be fined not less than 2,000 yuan and not more than 20,000 yuan, in addition.

Article 50 If agricultural products sold by an enterprise engaging in agricultural production or a professional farmers cooperative economic organization are under any of the circumstances listed in Items (1) through (3) or Item (5) of Article 33 of the present Law, the said entity shall be ordered to stop selling the products, replevy the sold agricultural products and make innocuous treatment over or destroy the illegally sold agricultural products under supervision; at the same time, its illegal income shall be confiscated and it shall be fined not less than 2,000 yuan and not more than 20,000 yuan, in addition.

If agricultural products sold by an enterprise engaging in sale of agricultural products are under any of the circumstances enumerated in the preceding paragraph, the said enterprise shall be punished in light of the preceding paragraph.

If on-sale agricultural products in an agricultural product wholesale market are under any of the circumstances listed in Paragraph 1 of this article, the agricultural products on illegal sale shall be dealt with in light of Paragraph 1 and the seller of the agricultural products shall be penalized in light of Paragraph 1.

If an agricultural product wholesale market violates Paragraph 1 of Article 37 of the present Law, it shall be ordered to make correction and be fined not less than 2,000 yuan and not more than 20,000 yuan.

Article 51 Whoever violates Article 32 of the present Law by imitating the quality marks on an agricultural product shall be ordered to make correction, his illegal proceeds shall be confiscated and he shall be fined not less than 2,000 yuan and not more than 20,000 yuan.

Article 52 The penalties prescribed in Article 44, Articles 47 through 49, Paragraphs 1 and 4 of Article 50 and Article 51 of the present Law shall be decided by the administrative department of agriculture of the people's government at the county level or above; while the penalties prescribed in Paragraph 2 and Paragraph 3 of Article 50 shall be decided by the administrative department for industry and commerce.

If any law has different provisions on an administrative penalty or the organ who has the power to make penalty, such provisions shall prevail, but the same illegal act shall not be penalized for twice or more.

Article 53 If someone violates the present Law and has committed a crime, investigations shall be conducted to determine his criminal liabilities in light of the law.

Article 54 If anyone who produces or sells the agricultural products enumerated in Article 33 of the present Law and causes any damages to the consumers, it shall undertake liabilities for compensation in light of the law.

If any on-sale agricultural product in an agricultural product wholesale market is under the circumstance prescribed in the preceding paragraph, the consumers may claim to the agricultural product wholesale market for compensation; if the producer or seller is held liable, the agricultural product wholesale market shall have the right to make recourse. The consumers concerned may also directly claim to the producer or seller of the agricultural products for compensation.

Chapter VIII Supplementary Provisions

Article 55 The administration on live pig slaughtering shall be carried out in light of the relevant provisions of the state.

Article 56 The present Law shall go into effect as of November 1, 2006.

Index